SpringerBriefs in Education

More information about this series at http://www.springer.com/series/8914

Elizabeth Warren · Jodie Miller

Mathematics at the Margins

Elizabeth Warren
Australian Catholic University
Virginia, QLD
Australia

Jodie Miller
Australian Catholic University
Virginia, QLD
Australia

ISSN 2211-1921 ISSN 2211-193X (electronic)
SpringerBriefs in Education
ISBN 978-981-10-0701-9 ISBN 978-981-10-0703-3 (eBook)
DOI 10.1007/978-981-10-0703-3

Library of Congress Control Number: 2016934200

Printed on acid-free paper

This Springer imprint is published by Springer Nature
The registered company is Springer Science+Business Media Singapore Pte Ltd.

This book is dedicated to the following people who made this project possible.

Eva deVries (Principal project officer), your expertise in early years mathematics contributed substantially to the success of the project. We thank you for your wealth of knowledge, endless work and ongoing commitment.

Danielle Armour (Indigenous research associate), thank you for your knowledge and contributions to the project. Your personal insights and guidance when working in, and with, Indigenous communities provided a cultural perspective that enriched the project. We wish you all the best and look forward to reading your Ph.D. thesis, exploring the experiences of Indigenous Teaching Assistants in school contexts.

Mari-Anne Wells (Project manager: Finance and data) and Denise Peck (Project manager: Administration), your efforts did not go unnoticed. Your endless hours of dedication to ensure everyone caught the right flights, had bags packed, and all the data were collected made this project a success. Thank you for all that you have done.

We also thank all the Research Assistants who have been involved over the period of the project.

Additionally, we thank principals, teachers, Indigenous teacher assistants and students who have participated in the RoleM project. Without you welcoming us into your school communities this would not have been possible. Thank you to many of you who invited the RoleM team into your lives; we have been privileged with life-long friendships.

Thank you for the ongoing generous support from community members and elders. Through your participation and commitment, RoleM was able to be implemented and have continued success.

Thank you Associate Professor Jackie, Associate Professor Bobbie and Dr. Jodie for your contributions to this book. Your insights into your own contexts are rich and invaluable.

Finally, the research reported in this book was supported by grants from Australian Research Council (LP100100154), and the Department of Education, Employment and Workplace Relations.

Preface

This book provides a positive story about teaching and learning mathematics in schools at the margins. While there are many challenges to overcome for both teachers and students in these contexts, effective learning can occur. Unique to this book, is the large longitudinal sample of students, teachers and Indigenous teacher aides that provide multiple perspectives with regard to the successful teaching and learning of mathematics (Foundation to Year 3). The purpose of sharing the journey of the teachers and students, is to enrich the readers' knowledge as to what quality teaching and learning can look like in mathematics in marginalised contexts. While this study is situated in Australian schools, the findings from the research have much to offer to both teachers and researchers in the international mathematics education community.

RoleM (Representations, oral language and engagement in Mathematics) began as a pilot study in 2008 with a small number of Aboriginal and Torres Strait Island schools. It was designed to bridge the educational gap in mathematics for young Indigenous students. In 2010, RoleM expanded to a 4-year longitudinal study focusing on a cohort of students from their first year of schooling, Foundation through to the completion of Year 3. A total of 16 schools were involved in the study. These schools were considered to be at the highest level of education risk in the Australian context. Over this 4-year period, the RoleM professional development model and the mathematics learning experiences were continually refined and adapted with the aim of maximising the quality of teaching and learning occurring in these contexts.

We wanted to share with readers the knowledge that was accumulated through this research, and contribute to the ongoing conversation about improving the teaching and learning of mathematics at the margins. Thus, the book is structured to tell the journey of the research that occurred over the four years. It presents the context and the frameworks that underpinned the study (Chap. 1) and critiques the international literature pertaining to quality learning and professional development (Chap. 2). Teacher and student data are presented separately (Chaps. 3 and 4) and then discussed together in Chap. 5. Chapter 6 reviews the main findings and presents implications for practice and research. Finally, the importance of maintaining the momentum and improving sustainable practices for teachers and students in these contexts is presented in Chap. 7. We would like to express our gratitude by thanking Associate Professor Jacqueline Ottman, Associate Professor Roberta Hunter, and Dr. Jodie Hunter who contributed to the final three chapters of the book. The insights and stories they have shared (Canadian and New Zealand perspectives) add great value to better understanding what is needed to improve teaching and learning mathematics in marginalised contexts.

Acknowledgements

Associate Professor Jacqueline Ottman contributes to the Canadian perspective in Chaps. 5, 6 and 7. Associate Professor Roberta Hunter and Dr. Jodie Hunter contribute to the New Zealand perspective in Chaps. 5, 6 and 7.

Contents

Chapter 1
Mathematics at the Margins

Abstract Schools classified as marginalized exhibit a complex cluster of factors, including parents who have low socio-economic status and low levels of education, and contexts where social networks are weak, there are few role models and in general there is a lack of opportunity. In the Australian context these schools tend to be in isolated geographical locations, and have large cohorts of students who either have English as their second language or are Indigenous. The problems that these schools face are universal. Teachers often struggle to work in these contexts, and students are at the greatest risk of not succeeding at school let alone mathematics. The focus of this book is to share the findings from a four-year longitudinal study *Representations, Oral Language and Engagement in Mathematics (RoleM)* that was situated in the most marginalized schools in Queensland, Australia. The participating students were in their first four years of school. The overall aim of the book is to share the journey of these teachers and students, and to draw out the dimensions that assisted these students to become successful learners of mathematics.

Defining Mathematics and Marginality

In this chapter we discuss/define (a) the terms 'mathematics' and 'marginality', (b) the intersection between the terms (mathematics and margins), and (c) present a conceptual framework that influences marginalized students' opportunities to learn. A brief synopsis of the other six chapters in the book is also provided. The notion of margins is deliberately chosen to underscore this book as it has many interpretations that are applicable to the research that this longitudinal study represents. For example, mathematics itself can marginalize students. Additionally, teachers assume that students at the margins of society will be unsuccessful in mathematics and hence do not afford them the opportunity to engage in high levels of mathematics. Thus, the notion of margins forms the foundation for the book's direction and subsequent discussions.

E. Warren and J. Miller, *Mathematics at the Margins*,
SpringerBriefs in Education, DOI 10.1007/978-981-10-0703-3_1

Marginalized Contexts

While marginalized contexts have traditionally been defined as neighborhoods with financial disadvantage, this is now seen to be too simplistic. 'Poverty line measures tend to belie the complexity and scope of disadvantage' (Price-Robertson 2011, p. 2). Marginalized communities are disadvantaged in all aspects of life and result from the interplay between unemployment, racial isolation, social exclusion, low educational levels, financial dependence, and drug and alcohol abuse (Berman and Paradies 2010; Price-Robertson 2011; Sampson 2000). They exhibit weak social networks, poor role models and relative lack of opportunity (Edwards 2005; Vinson et al. 2007). These types of neighborhoods are a worldwide phenomenon, existing in remote, rural, and urban contexts. Additionally, they largely consist of minority groups such as African-Americans, Latinos, Indigenous people, ethnic minorities, and Pacifica, as these groups are more likely to cluster in marginalized contexts (Catsambis and Beveridge 2001; Massey and Denton 1993).

Schools in marginalized contexts share five common traits:

- Students exhibit educational poverty. They tend to be situated in the lowest levels of a variety of performance measures (e.g., national and international tests of literacy and numeracy).
- Students often do not have the official language of instruction (UNESCO 2014).
- There is a high staff turnover, and schools in these contexts experience difficulties in attracting and retaining high quality teachers (Borman and Dowling 2008; Lyons et al. 2006).
- These schools commonly possess poor management and poor performance measures (Lupton 2004).
- The teachers they tend to attract are inexperienced and lack a commitment to teaching in these contexts (Duru-Bellat 2009; Heslop 2011; Mills and Gale 2010).

In Queensland, the second largest state in Australia and the site of the RoleM project, numeracy learning in the early years is even more problematic. Thorpe and colleagues (2004), in a large study investigating learning experiences and teaching practices prior to Year 1, found that many students made negative progress in their understanding of basic numeracy concepts. This is consistent with the findings of Denton and West (2002), in their large study of 20,000 students, who found that students from low-income families gained little understanding of mathematics in their first year of schooling. For students from marginalized contexts this is even more problematic, as they enter school with little knowledge of Western mathematics (Field et al. 2007). In addition, students attending schools in marginalized contexts often come from backgrounds where the home language is commonly not Standard Australian English. This is often challenging, as Australia is a mono-linguistic culture where the vast majority of people only speak English, and English is the language used in nearly every school. Thus, students who exhibit low English language proficiency and low socio-economic status are at most risk of poor schooling outcomes (Goldenberg 2008; UNESCO 2014). But schools and teachers situated in these contexts also inadvertently contribute to these poor outcomes.

Internationally, students in marginalized contexts appear to be disproportionally taught by under-qualified teachers (Borman and Kimball 2013). In Australia, very few teachers entering these contexts feel prepared both educationally and culturally by their pre-service training to effectively teach in these contexts (MCEECDYA 2011; White and Reid 2008), and their interactions with people of other ethnic backgrounds or different social classes prior to their entry have been limited (Allard and Santoro 2004). Additionally, in Queensland many of these schools are in geographically isolated locations, either in rural, remote or very remote areas of the state or at the margins of metropolitan cities. Thus, these teachers who are predominantly from metropolitan areas feel isolated, both professionally and socially. Due to the high staff turnover, there are few resources or mentors to assist them to be effective, with many feeling unable to create highly effectual instructional programs (Kent 2004). A consequence of this is that there is a propensity to create highly structured classes with a heavy reliance on the use of worksheets (Hewitson 2007), and a classroom environment that gives no credence to the backgrounds or cultural diversity of the participating students.

Dominating the literature pertaining to marginalized contexts are the metaphors of educational gap, education debt or deficits, and closing these gaps. The research findings from these settings tend to document the deficits in terms of quantifiable differences rather than focus on how to redress these issues (Gutiérrez and Dixon-Román 2011). While many of these teachers recognize a need for changed practices to enable these students to learn, they have little knowledge of what these practices would look like (Jorgensen et al. 2010).

There is also a large body of research in the mathematics field that strongly supports repositioning the notion of what constitutes mathematics for these students by identifying and studying mathematics that is embedded in their existing culture: an ethnographic approach to teaching mathematics (Matang and Owens 2004). With regard to the Indigenous Australian students, while these approaches have proven to assist students to engage in mathematics, there are instances 'where the infiltration of anthropological schemes into Indigenous educational practices have been counterproductive, giving the anthropological discourse primacy over the educational context' (Nakata 2003, p. 9). Nakata also suggests that this cultural difference scheme can serve to provide a convenient explanation for students' failure.

Another approach to teaching mathematics is to draw on real world problems that are seen as applicable to these students (Boaler 1993). This approach is underpinned by the premise that this results in intellectually challenging and engaging work which in turn results in improved student outcomes (Boaler 1997; Herrington et al. 2008). Jurdak (2009) purports that these efforts, while producing positive measureable outcomes, may in fact result in different levels of mathematical literacy, and consequently increase the potentiality of further marginalizing the marginalized. This resonates with the stance that, in many instances, school mathematics has strayed too far from its disciplinary referents (Ball 1993).

Within the Australian community there have been calls for taking a more balanced perspective when considering school learning, and recognizing that different ways of knowing are legitimate. One way does not need to be at the expense of another. This is reflected in the literature pertaining to two-way learning (Harris 1994; Sarra 2006), orbiting between two worlds (Pearson 2009), or both ways strong (Barnhardt 2008; Jorgensen and Lowrie 2013). History has shown that while it has been recognized that different cultures hold different knowledges, little is known about how to utilize these in the classroom. In the case of mathematics, the dominant knowledge within the Australian context is Western mathematics. Recently there has been a push for students keeping their culture identity, receiving a quality education (Pearson 2009) and, sociopolitical consciousness (understanding the structural constraints that inhibit some groups from advancing) (Ladson-Billings 2000). Thus mathematics at the margins is about maintaining a fluency in students' culture concurrently with supporting them to excel in Western mathematics; an attempt to intertwine both dimensions. Accordingly, the focus has moved beyond students, their teachers and mathematics to include teacher assistants, particularly teacher assistants with high levels of cultural knowledge.

Mathematics in These Contexts

Defining Mathematics

Although it is acknowledged that mathematics is a fundamental construct required for operating successfully in society, there is debate with regard to what mathematics is, especially within the elementary school. This debate ranges from learning basic facts and skills to engaging in higher order thinking. The first is situated in the notion of being computationally ready to answer problems, and the second reflects the philosophical stance that mathematics is about its creation. Thus many current curriculum documents purport that mathematics as a subject is ever evolving and that mathematical ideas have evolved over thousands of years across all cultures (e.g., Australian Mathematics Curriculum, and others). From this perspective the emphasis in school mathematics is seen as sharing with students an appreciation of its elegance and power, and developing in students the thinking that is at the heart of its creation: mathematical reasoning, analytical thought, and the ability to pose and solve problems (e.g., NCTM Standards 2000). Underlying this philosophical stance is the need to be competent at particular conceptual understandings of mathematics with a focus on skills and procedures. The content is determined by the curricula (Hewson 2005), and together with the move to develop students' ability to think mathematically constitute the *envisioned curriculum* (Trafton et al. 2001).

Mathematics as a Marginalizing Practice

Both teachers and students can feel marginalized as they teach and learn mathematics. First, it is well recognized that school mathematics can act as a social filter to reproduce the status quo (Ernest 2007; Schoenfeld 2002; Zevenbergen 2001;), with many students opting out of mathematics in their later years. It is also used as an intellectual filter to select students to participate in higher levels of mathematics and science; subjects that are fundamental to many highly paid employment opportunities (Fullarton et al. 2003; Hewson 2005; Lamb and McKenzie 2001). Second, teachers in the early years are often under-confident at teaching mathematics and thus feel marginalized by mathematics as a subject they are required to teach. A direct consequence of this is their predominant focus on literacy in the early years at the expense of teaching mathematics (Department of Education and Early Childhood Development 2009; Sun Lee and Ginsburg 2007). Third, teachers situated in disadvantaged contexts often hold very stereotypical views about their students' ability to learn mathematics, and thus offer a dumbed down curriculum to their students (Matthews et al. 2005). This serves to further marginalize these students as it has been shown that there is direct correlation between strong foundations in the early years and strong mathematical outcomes in the later years of schooling (Denton and West 2002; Lerkkanen et al. 2005; Stevenson and Newman 1986).

School Mathematics

While the mathematics curriculum delineates what is envisioned (intended) to happen in the mathematics classroom, the teacher mediates what actually happens (Porter 1989). Teachers make decisions about how much time to spend on teaching mathematics, what topics to teach, in what order to teach them, and the role of teacher assistants in the learning process. Teachers' knowledge of mathematics and the pedagogy of teaching mathematics, and their knowledge of the learner and their situations influence these choices. Their choices are also influenced by the topics that teachers are personally interested in Ball et al. (2005). Thus, what is envisioned for the classroom and what is enacted in the classroom do not necessarily align. Furthermore, meaning and learning rely on a sense of confidence in doing mathematics, a belief about its relevance and a desire to know about mathematics (Otte 2005). Hence the beliefs teachers have about mathematics and the confidence they have in teaching mathematics, both impact on the reality of what happens in the classroom and what is implemented (the *enacted curriculum*).

RoleM in Marginalized Contexts

Representations Oral Language and Engagement in Mathematics (RoleM) was a four-year longitudinal project situated in the first four years of schooling (Foundation to Year 3). All participating schools were from marginalized contexts. Underpinning the project design is the philosophical stance that a key dimension of mathematics is the ability to effectively communicate about mathematical concepts. This communication consists of the language of mathematics together with an array of representations. For many students from marginalized contexts both of these have proved to be barriers to their effective engagement in mathematics (Gutiérrez 2002; Matthews et al. 2003). Not only are representations essential to students' understanding of mathematics, but also 'seeing' concepts in a variety of different representations can deepen their understanding (Arcavi 2003; Goldin and Shteingold 2001; Pape and Tchoshanov 2001). In addition, it was acknowledged that teachers in these contexts needed to be fully supported as they transition into these communities. Thus the decision was made to create research-based, proven learning activities and provide teachers with collegial mentoring as they trialed these activities in the setting of their classroom.

The project design consisted of iterative cycles comprising four components: design, trial, evaluation and modification. For each year of the project, 35 learning activities were developed and trialed. In the trial stage, these activities were shared with participating teachers, and members of the team worked with teachers in the classroom to trial the activities. Data were gathered with regard to their effectiveness, and appropriate modifications were made.

The participants were drawn from 16 school communities across Queensland, one of the largest states in Australia. Queensland's population is mainly situated along its extensive eastern coastline, with the population clustered in metropolitan or large provincial towns. The sample was drawn from three different locations across Queensland: metropolitan/provincial, remote, and very remote. Very remote schools are in geographical locations that are spatially very remote from provincial and metropolitan areas (Jones 2004).

Metropolitan/provincial geographic locations include cities with populations of 100,000 or more situated in non-remote areas of Queensland. All are easily accessible by road and air, with populations representative of a wide range of socio-economic groups. These locations provide a range of education opportunities for their children, including higher education courses offered by local universities and vocational education providers.

Remote and very remote locations, due to their distance and isolation from large centers, have very restricted access to goods and services and opportunities for social interaction. They often comprise vast regions of uninhabited areas along with small isolated rural towns. Due to the large distances separating many of these locations, the delivery of education and health care services is extremely difficult. In addition, the proportion of Indigenous people increase with increasing remoteness (AIHW 2011). The more remote the location, the more the education outcomes lag

Table 1.1 Participating schools across the four years with the frequency of students and teachers

School location	Schools	Students			Teachers
		Indigenous	ESL	Other	
Metropolitan	7	210	517	348	89
Remote	5	318	2	197	45
Very remote	4	132	0	14	20
Total	16	660	519	559	154

behind those of other Australians (ABS 2008). For us, access to the participating remote and very remote schools comprised travel of up to eight hours duration, including a four-hour plane flight followed by a four-hour car journey.

Table 1.1 summarizes the number of schools in each geographical location, together with the frequency of students that participated in the data gathering. Students are classified according to whether they are Indigenous Australian students, English as a Second Language (ESL) students, or other students. Students in the other category were predominantly white Australian students. In the Australian context, Indigenous students are not generally classified as ESL students. While it is recognized that for many Indigenous students English is their second language, the ESL classification refers to students who recently arrived in Australia and do not speak English.

A total of 1738 students fully participated in the data collection, with the distribution of the total students being evenly spread across the three groups (Indigenous, ESL and other).

The Conceptual Framework that Underpinned RoleM

It is well documented that students in marginalized contexts continue to struggle engaging with and learning Western mathematics (e.g., Clarkson 1992; COAG 2008; Jorgensen 2010). It is also recognized that quality teachers (how teachers enact the mathematics curriculum in their classrooms) can make a difference to their students' scholastic outcomes (Hattie 2009; Smart et al. 2008). However, how teachers interact with, draw on, and are influenced by mathematics resources and instructional guidelines in classrooms is a contentious issue with practices ranging from close adherence to almost ignorance of what is available and written (Chavez 2003). Three dimensions that are recognized as influencing their choices are teachers' pedagogical content knowledge, subject matter knowledge (Shulman 1986), and their educational goals for their students (Wilson and Goldenberg 1998). In marginalized contexts we contest that there are three other dimensions that also influence what teachers enact: their knowledge of the context/culture that their students are situated in and how they best learn; the confidence that they have in teaching mathematics to these students; and their beliefs about the capability of

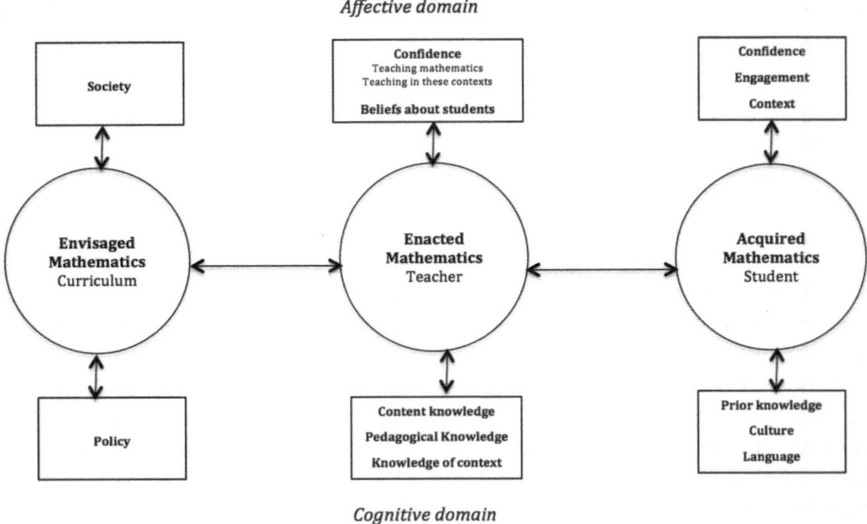

Fig. 1.1 Model depicting the relationship between teaching and learning and the dimensions that influence its effectiveness

these students to learn mathematics. Hence, a key to making a difference entails a focus on teachers' enacted curriculum, what actually occurs in the context of the classroom.

The conceptual framework illustrated in Fig. 1.1 emerged from the research findings with regard to teaching and learning mathematics in marginalized contexts (Boaler and Staples 2008; Gutstein 2003; Gutiérrez 2002). The predominant 'players' in this framework are the teachers and their students. Within the mathematics classroom in marginalized contexts, the learning and teaching of mathematics is influenced by a number of key elements or dimensions. Fundamentally, the mathematics that students are to learn is dictated by the mathematics curriculum – the *envisioned curriculum* – an official set of learning outcomes delineated by the national and state frameworks. Both the teachers and students heavily influence what actually happens in these classrooms: the *enacted curriculum*. As with all classrooms, teachers' own knowledge, beliefs and experiences, in conjunction with students' prior knowledge and willingness to learn, influence the teaching learning process. An added dimension to this interplay is the context and culture in which the teaching and learning is occurring. The teachers themselves are often the product of white middle class backgrounds with little experiences of these marginalized contexts, and students enter the classroom from culturally diverse backgrounds that do not necessarily make them 'school ready' from the teachers' perspective. We contend that there are a series of strategies to help ensure that teaching and learning mathematics is effective. These rely on simultaneously supporting both teachers and students, and building on what both parties bring to the table. Primarily teachers in these contexts feel unprepared, unsupported and under-resourced, and students enter these schools without a strong background in the use of English – the predominant language of classroom discourse – or strong foundations in Western mathematics. We also acknowledge that both players do have existing strengths. Teachers desire to be effective and students enter school with a readiness to learn (Denton and West 2002).

Consistent with our conceptual framework, the primary question that drove the RoleM project was:

What is the interplay between effective teaching and effective learning in these contexts?

The sub-questions were:

How do teachers' dispositions towards teaching mathematics influence their classroom practice in these contexts?

How does teachers' mathematics content knowledge and pedagogical knowledge influence their classroom practices in these contexts?

How does professional learning best support the development of this interplay?

Overview of the Book

The book comprises seven chapters. A brief overview of subsequent chapters is:

Chapter 2 Being at the margins ≠ being unsuccessful at mathematics
This chapter critiques the international literature pertaining to teaching mathematics to early years students from marginalized contexts, identifies the actions that assist these students learn mathematics, and draws implications with regard to providing quality professional learning for these teachers.

Chapter 3 Mathematics and marginality
This chapter explores the issues of mathematics as a marginalizing tool and how teachers in these contexts typically deliver mathematics to their students. It shares the teacher data and examines their professional learning progress over the four-year period of the RoleM project. Illustrative examples are included to evidence the types of teaching actions that make a difference.

Chapter 4 Marginality and mathematics
This chapter explores the issue of marginality in terms of the context in which the participating students in RoleM live, their cultural and educational background and the impact this has on their learning of mathematics. It shares illustrative examples of how these students' engagement with and learning of mathematics evolved over the four-year period.

Chapter 5 Crossing the divide
This chapter discusses the teacher and student data in terms of the literature. Recommendations are drawn for curriculum developers, education providers, and educational policy.

Chapter 6 Redressing the imbalance
This chapter outlines recommendations in regard to the learning of mathematics, with a particular focus on assisting young ESL and Indigenous students to effectively engage in Western mathematics.

Chapter 7 Maintaining the momentum
This chapter discusses the importance of maintain the momentum and gaining suability for students and teachers in marginalized contexts. Conclusions are drawn, and actions delineated that ensure a smooth transition for these young students throughout their first four years of school.

At the conclusion of Chaps. 5, 6 and 7, Canadian and New Zealand researchers reflect on the issues that emerged from the RoleM project and discuss these issues with regard to their particular marginalized contexts.

References

Allard, A., & Santoro, N. (2004). Making sense of difference? Teaching identities in postmodern contexts. In P. Jeffery (Ed.), *Doing The Public Good: Positioning Educational Research (AARE 2004 International Education Research Conference Proceedings)* (pp. 1–20). Coldstream, VIC: Australian Association for Research in Education.

Arcavi, A. (2003). The role of visual representations in the learning of mathematics. *Educational studies in mathematics, 52*(3), 215–241.

Australian Bureau of Statistics. (2008). *Australian social trends: Education across Australia* (Cat. no. 4102.0). Canberra, Australia: ABS.

Australian Institute of Health and Welfare (AIHW). (2011). *The heath and welfare of Australia's Aboriginal and Torres Strait Islander people*. Canberra, ACT: Author.

Ball, D. L. (1993). With an Eye on the mathematical horizon: Dilemmas of teaching elementary school mathematics. *The Elementary School Journal, 93*(4), 373–397.

Ball, D. L., Hill, H. C., & Bass, H. (2005). Knowing Mathematics for Teaching. *American Educator, 29*(1), 14–17, 20-22, 43–46. doi:10.1016/j.cedpsych.2006.02.001

Barnhardt, R. (2008). Creating a place for indigenous knowledge in education: The Alaskan native knowledge network. In D. A. Gruenewald & G. A. Smith (Eds.), *Place-based education in the global age: Local diversity* (pp. 113–134). New York, NY: Taylor Francis.

Berman, G., & Paradies, Y. (2010). Racism, disadvantage and multiculturalism: Towards effective anti-racist praxis. *Ethnic and Racial Studies, 33*(2), 214–232.

Boaler, J. (1993). Encouraging the transfer of 'school' mathematics to the 'real world' through the integration of process and content, context and culture. *Educational Studies in Mathematics, 25* (4), 341–373.

Boaler, J. (1997). Equity, Empowerment and Different Ways of Knowing. *Mathematics Education Research Journal, 9*(3), 325–42.

Boaler, J., & Staples, M. (2008). Creating mathematical futures through an equitable teaching approach: The case of Railside School. *The Teachers College Record, 110*(3), 608–645.

Borman, G. D., & Dowling, N. M. (2008). Teacher attrition and retention: A meta-analytic and narrative review of the research. *Review of Educational Research, 78*(3), 367–409.

Borman, G. D., & Kimball, S. M. (2013). Teacher quality and educational equality: Do teachers with higher standards- ratings close student achievement gaps? *The Elementary School Journal, 106*(1), 3–20.

Catsambis, S., & Beveridge, A. A. (2001). Does neighborhood matter? Family, neighborhood, and school influences on eighth-grade mathematics achievement. *Sociological Focus, 34*(4), 435–457.

Chavez, O. L. (2003). *From the textbook to the enacted curriculum*. Unpublished doctoral dissertation, University of Missouri, Columbia, MO.

Clarkson, P. C. (1992). Language and mathematics: A comparison of bilingual and monolingual students of mathematics. *Educational Studies in Mathematics, 23*(4), 417–429.

Council of Australian Governments (COAG). (2008). National numeracy review report. Canberra, ACT. Retrieved from http://www.coag.gov.au/sites/default/files/national_numeracy_review.pdf

Denton, K., & West, J. (2002). *Children's reading and mathematics achievement in kindergarten and first grade*. Washington: National Center for Education Statistics, US Department of Education.

Department of Education, Employment, and Early Childhood Development. (2009). *Numeracy in practice: Teaching, learning and using mathematics*. Melbourne, Vic: Education Policy and Research Division.

Duru-Bellat, M. (2009). *Access to education: what are the inequalities in France today?*. Report: Background paper for EFA Global Monitoring. 2010.

Edwards, B. (2005). Does it take a village? An investigation of neighbourhood effects on Australian children's development. *Family Matters, 72*, 36–43.

Ernest, P. (2007). Why social justice. *Philosophy of Mathematics Education Journal, 21*. Retrieved from http://people.exeter.ac.uk/PErnest/pome21/index.htm

Field, S., Kuczera, M., & Pont, B. (2007). *No More Failures: Ten Steps to Equity in Education*. Paris: OECD.

Fullarton, S., Walker, M., Ainley, J., & Hillman, K. (2003). *Patterns of participation in Year 12*. Melbourne, Vic.: ACER.

Goldenberg, C. (2008). Teaching English language learners. *American Educator*, 8 – 44. Retrieved from http://old.dmps.k12.ia.us/programs/7researcharticle.pdf

Goldin, G., & Shteingold, N. (2001). Systems of representation and the development of mathematical concepts. In A. A. Cuoco & F. R. Curcio (Eds.), *The role of representation in school mathematics* (pp. 1–23). Boston, MA: NCTM.

Gutstein, E. (2003). Teaching and learning mathematics for social justice in an urban, Latino school. *Journal for Research in Mathematics Education*, 37-73.

Gutiérrez, R. (2002). Enabling the practice of mathematics teachers in context: Toward a new equity research agenda. *Mathematical Thinking and Learning, 4*(2–3), 145–187.

Gutiérrez, R., & Dixon-Román, E. (2011). Beyond gap gazing: How can thinking about education comprehensively help us (re)envision mathematics education? In B. Atweh, M. Graven, W. Secada, & P. Valero (Eds.), *Mapping equity and quality in mathematics education* (pp. 21–34). New York: Springer.

Harris, S. (1994). 'Soft' and 'hard' domain theory for bicultural education in Indigenous groups. *Peabody Journal of Education, 69*(2), 140–153.

Hattie, J. A. (2009). *Visible learning: A synthesis of 800+ meta-analyses on achievement*. Abingdon: Routledge.

Heslop, J. (2011). Living and teaching in Aboriginal communities. In Q. Beresford & G. Partington (Eds.), *Reform and resistance in Aboriginal education* (pp. 208–238). Perth: University of Western Australia Press.

Herrington, J., Reeves, T., & Oliver, R. (2008). Authentic Learning Environments. In M. Spector, M. Merril, J. Elen, & M. Bishop (Eds.), *Handbook of Research in Educational Communications and Technology* (pp. 425–428). New York: Springer.

Hewitson, R. (2007). Climbing the Educational Mountain: A Metaphor for Real Culture Change for Indigenous Students in Remote Schools. *Australian Journal of Indigenous Education, 36*, 6–20.

Hewson, G. (2005). The meaning of mathematics. In J. Kilpatrick, C. Hoyles, & O. Scovsmose (Eds.), *Meaning of mathematics education* (pp. 17–38). New York: Springer.

Jones, R. (2004). Geolocation questions and coding index. A technical report submitted to the MCEETYA Performance Measurement and Reporting Taskforce. Retrieved from www.mceetya.edu.au/mceetya/default.asp?id=11968.

Jorgensen, R. (2010). Structured failing: Reshaping a mathematical future for marginalized learners. L. Sparrow, B. Kissane, & C. Hurst (Eds.), *Shaping the future of mathematics education: Proceedings of the 33rd annual conference of the Mathematics Education Research Group of Australasia* (pp. 26-35). Fremantle, WA: MERGA.

Jorgensen, R., & Lowrie, T. (2013). Both ways strong: Using digital games to engage Aboriginal learners. *International Journal of Inclusive Education, 17*(2), 130–142.

Jorgensen, R., Grootenboer, P., Niesche, R., & Lerman, S. (2010). Challenges for teacher education: The mismatch between beliefs and practice in remote Indigenous contexts. *Asia-Pacific Journal of Teacher Education, 38*(2), 161–175.

Jurdak, M. (2009). *Toward Equity in Quality in Mathematics Education*. Boston, MA: Springer.

Kent, A. M. (2004). Improving teacher quality through professional development. *Education, 124* (3), 427–435.

Ladson-Billings, G. (2000). Culturally relevant pedagogy in African-centred schools: Possibilities for progressive educational reform. In D. Pollard & C. Ajirotutu (Eds.), *African-centered schooling theory and practice* (pp. 188–198). London: Bergin & Garvey.

Lamb, S., & McKenzie, P. (2001). *Patterns of Success and Failure in the Transition from School to Work in Australia, LSAY Report 18*. Melbourne, Vic.: ACER.

Lerkkanen, M. K., Rasku-Puttonen, H., Aunola, K., & Nurmi, J. E. (2005). Mathematical performance predicts progress in reading comprehension among 7-year olds. *European journal of psychology of education, 20*(2), 121–137.

Lupton, R. (2004). Schools in Disadvantaged Areas : Recognising Context and Raising Quality. Retrieved from http://sticerd.lse.ac.uk/dps/case/cp/CASEpaper76.pdf

Lyons, T., Cooksey, R., Panizzon, D., Parnell, A., & Pegg, J. (2006). *Science, ICT and mathematics education in rural and regional Australia: The SiMERR national survey*. Canberra: Department of Education, Science and Training.

Massey, D. S., & Denton, N. A. (1993). *American apartheid: Segregation and the making of the underclass*. Boston: Harvard University Press.

Matang, R., & Owens, K. (2004). Rich transitions from Indigenous counting systems to English arithmetic strategies: Implications for mathematics education in Papua New Guinea. In F. Favilli (Ed.), *Proceedings of the 10th International Congress on Mathematical Education, Copenhagen, Denmark, Discussion Group 15, Ethnomathematics: Ethnomathematics and mathematics education* (pp. 107–117). Pisa, Italy: ICME.

Matthews, S., Howard, P., & Perry, B. (2003). Working together to enhance Australian Aboriginal students' mathematics learning. In L. Bragg, C. Campbell, G. Herbert, & J. Mousley (Eds.), *Proceedings of the 26th Annual Conference of the Mathematics Education Research Group of Australasia (MERGA 26)* (pp. 9–28). Geelong, VIC: MERGA.

Matthews, C., Watego, L., Cooper, T. J., & Baturo, A. R. (2005). Does mathematics education in Australia devalue Indigenous culture? Indigenous perspectives and non-Indigenous reflections. In P. Clarkson, A. Downtown, D. Gronn, M. Horne, A. McDonough, R. Pierce, & A. Roche (Eds.), *Proceedings of the 28th Annual Conference of the Mathematics Education Research Group of Australasia* (pp. 513–520). Melbourne, Vic.: University of Melbourne.

Ministerial Council for Education Early Childhood Development and Youth Affairs (MCEECDYA). (2011). *Aboriginal and Torres Strait Islander Education Action Plan 2010–2014*. Retrieved from http://www.mceecdya.edu.au/verve/_resources/A10-0945_IEAP_web_version_final2.pdf

Mills, C., & Gale, T. (2010). Schooling in disadvantaged communities. *Journal of Social Inclusion, 1*(2), 181–182.

Nakata, M. (2003). Some thoughts on literacy issues in Indigenous contexts. *Australian Journal of Indigenous Education, 31*, 7–15.

National Council of Teachers of Mathematics (NCTM). (2000). *Principles and standards for school mathematics*. Reston, VA: Author.

Otte, M. (2005). Meaning and Mathematics. In *Meaning in Mathematics Education* (pp. 231-260). Springer US.

Pape, S. J., & Tchoshanov, M. A. (2001). The role of representation (s) in developing mathematical understanding. *Theory into practice, 40*(2), 118–127.

Pearson, N. (2009). Radical hope: Education and equality in Australia. *Quarterly Essay, 35*, 1–105.

Porter, A. (1989). A Curriculum out of Balance The Case of Elementary School Mathematics. *Educational Researcher, 18*(5), 9–15.

Price-Robertson, R. (2011). *CAFCA resource sheet Understanding the issues, overcoming the problem* (pp. 1–10). Retrieved from https://www3.aifs.gov.au/cfca/sites/default/files/publication-documents/rs2.pdf

Sampson, R. J. (2000). The neighborhood context of investing in children: facilitating mechanisms and undermining risks. In S. Danziger & J. Waldfogel (Eds.), *Securing the future: investing in children from birth to college* (pp. 205–227). New York: Russell Sage Foundation.

Sarra, C. (2006). *Don't blame me.* Retrieved from http://www.eqa.edu.au/site/dontblameme.html

Shulman, L. S. (1986). Those who understand: Knowledge growth in teaching. *Educational Researcher, 15*(2), 4–14.

Schoenfeld, A. H. (2002). Making mathematics work for all children: Issues of standards, testing, and equity. *Educational researcher, 31*(1), 13–25.

Smart, D., Sanson, A., Baxter, J., Edwards, B., & Hayes, A. (2008). *Home-to-school transitions for financially disadvantaged children: Final report*. Sydney: The Smith Family.

Stevenson, H. W., & Newman, R. S. (1986). Long-term prediction of achievement and attitudes in mathematics and reading. *Child development, 57*, 646–659.

Sun Lee, J., & Ginsburg, H. P. (2007). Preschool teachers' beliefs about appropriate early literacy and mathematics education for low-and middle-socioeconomic status children. *Early Education and Development, 18*(1), 111–143.

Thorpe, K., Tayler, C., Bridgstock, R., Grieshaber, S., Skoien, P., Dany, S., & Petriwsky, A. (2004). Preparing for school: re-port on Queensland preparing for school trials. Retrieved 7th November 2008, from http://eprints.qut.edu.au/10192/1/10192.pdf.

Trafton, P. R., Reys, B. J., & Wasman, D. G. (2001). Standards-based mathematics curriculum materials: A phrase in search of a definition. *The Phi Delta Kappan, 83*(3), 259–264.

UNESCO. (2014). *Chapter 3 Reaching the marginalized. Teaching and learning: Achieving quality for all* (pp. 133–213). Retrieved from http://publishing.unesco.org/details.aspx?Code_Livre=5018

Vinson, T., Rawsthorne, M., & Cooper, B. (2007). *Dropping off the edge : The distribution of disadvantage in Australia*. Richmond, VIC: Jesuit Social Services.

White, S., & Reid, J. (2008). Placing teachers? Sustaining rural schooling through place consciousness in Teacher Education. *Journal of Research in Rural Education, 23*(7), 1–11.

Wilson, M., & Goldenberg, M. (1998). Some conceptions are difficult to change: one middle school mathematics teacher's struggle. *Journal of Mathematics Teacher Education, 1*, 269–293.

Zevenbergen, R. L. (2001). Literacy learning : the middle years. *Literacy in the Middle Years, 9*(2), 21–28.

Chapter 2
Being at the Margins ≠ Being Unsuccessful at Mathematics

Abstract Students at the margins are disadvantaged as they enter school, and the educational gap between disadvantaged and advantaged students widens as they progress through school. We would suggest that a primary cause for this occurring is how the envisaged mathematics curriculum is enacted in these classroom contexts. Thus this chapter is organized under the two major dimensions that are purported to influence the enactment of mathematics teaching: teachers' affective domain and their cognitive domain. This chapter critiques the literature relating to these domains and identifies particular elements that assist this educational gap being 'closed'. It also shares the theoretical constructs that underpinned the selection and development of materials for the classroom and the development of the RoleM professional learning model utilized across the four years of the project.

Affective Domain for Enacting Mathematics

Dispositions for Teaching Mathematics

The affective domain for teaching mathematics consists of three predominant multidimensional constructs: beliefs, attitudes, and emotions (Ma and Kishor 1997; McLeod 1992; Philipp 2007). These are commonly used to describe a wide spectrum of teachers' affective responses to mathematics. The three constructs represent varying levels of affective involvement, intensity and stability (Grootenboer et al. 2008). Beliefs are considered to be notions held by individuals that they consider as true (Grootenboer et al. 2008). Beliefs are more stable, cognitively related, and develop over a long period of time. Attitudes are learnt and are evident in responses to situations. For example, attitudes are responses such as confidence and anxiety, engaging and avoiding, and liking and disliking mathematics (Beswick et al. 2006; Ernest 1988; Ma and Kishor 1997). Recently there has been a greater emphasis placed on the role that emotions play in the learning of mathematics. DeBellis and

E. Warren and J. Miller, *Mathematics at the Margins*,
SpringerBriefs in Education, DOI 10.1007/978-981-10-0703-3_2

Goldin (2006) see emotions or human affect operating as an internal representational systems parallel to cognitive systems as students learn mathematics. They can be seen as two complimentary aspects of mind (Hannula 2002, p. 27). Thus states of emotion carry meaning for the individual. For example, a student's feeling of frustration may be internally represented as failure, and result in avoidance strategies occurring. Emotions can thus empower or disempower students as they learn. In a problem solving contexts an understanding of this emotional dimension can be crucial to success as disempowering emotions can hamper performance. Emotions or feelings are affective responses to particular contexts or events. As such they are transitory and unstable in nature (Schuck and Grootenboer 2004).

Teachers' beliefs about students change from context to context. This is problematic for students from marginalized areas, as many teachers believe they are incapable of learning. In a remote Australian Indigenous community, Jorgensen and colleagues (2010) found that there was a mismatch between the teachers' beliefs and what was delivered to students in these marginalized contexts. Teachers often espoused beliefs about good teaching practices, but did not incorporate them into classroom teaching practices. Commonly, teachers provided lessons based on skill and drill learning experiences (Baturo et al. 2008; Jorgensen et al. 2010). These findings align with the beliefs teachers hold in relation to Indigenous students' mathematical ability (Matthews et al. 2005). Thus it is imperative to move teachers towards a mindset of quality teaching and learning for all, regardless of context.

Teachers' beliefs and attitudes about teaching and learning affect their teaching practices (Ambrose 2004; Kagan 1992). Teachers have a profound influence on the learning environment, and this is partially determined by their self-concept (beliefs about themselves) and the confidence they have in teaching mathematics (Rohrkemper 1986). Under-confident teachers tend to create learning environments that are unsupportive of learning. While effective teachers' tend to create learning environments where students are encouraged to be enthusiastic and enjoy mathematics (McLeod 1992). In the literature, self-concept is often associated with the term self-efficacy; that is, teachers' belief about their capability to bring about desired outcomes in their students' engagement and learning (McLeod 1992; Stipek et al. 2001).

Teachers' self-efficacy impacts on how they feel, think, motivate themselves, behave and perform (Bandura 1992). Those with low self-efficacy visualize failure scenarios and labor about the many things that can go wrong. Even though they may have the knowledge and skills to create effective learning environments for their students, such teachers may find it difficult to use these under taxing conditions. As teachers gain progressive mastery over difficult contexts, their own performance is enhanced (Bandura 1991). Thus, the task of creating effective learning environments for students in disadvantaged contexts rests heavily on the self-efficacy of the teacher and beliefs in their ability to effectively teach their students (instructional efficacy).

Teachers' self-efficacy is closely related to their instructional efficacy. When teachers' instructional efficacy is high they devote more time to academic learning, assist students who are having difficulties, and praise students on their accomplishments. If their instructional efficacy is low, they give up more easily on students, students are held accountable for their failures, and teachers generally undermine their students' abilities, efficacy and cognitive development. Within marginalized contexts, settings with higher proportions of students from low socio-economic backgrounds and higher levels of student turnover and absenteeism, teachers' instructional efficacy is predominantly low (Ashton and Webb 1986).

The confidence teachers possess and year level in which they teach also influences their instructional efficacy. For example, early years teachers are often under-confident when teaching mathematics. The literature identifies two areas where many of these teachers lack confidence: their own understandings of mathematics; and, their ability to teach mathematics. This lack of confidence has the potential to impact on (a) the types of mathematical tasks these teachers utilize, and (b) the instructional decisions they make as they teach mathematics (Chick and Beswick 2013; Sullivan et al. 2009), resulting in low levels of instructional efficacy. Clarke and colleagues (2009) hypothesize that teachers with low instructional efficacy choose an instructional pathway that consists of guiding students incrementally through a set of skills. This limits the opportunities for students to engage in high levels of mathematical thinking (Staub and Stern 2002; Stein and Lane 1996).

Thus, the affective issues that influence teachers in these marginalized contexts are (a) their own self efficacy with regard to how they feel about working in these unfamiliar contexts and how they think they will perform, (b) their beliefs about whether these students are capable of learning high levels of mathematics, and (c) their own instructional efficacy; that is, their belief in their capability to create effective learning experience for these students.

Cognitive Domain for Enacting Mathematics Teaching

Knowledge for Teaching Mathematics

In research there is shared understanding that a distinct construct, *knowledge for teaching mathematics,* underscores quality student learning. Historically, what this means has changed over the last 40 years. There has been a movement from knowledge being conceived as a static entity (e.g., Ernest 1989; Shulman 1986) to knowledge as a dynamic entity (e.g., Rowland et al. 2005), with the former categorizing knowledge into different types and the latter including the capacity to know how to act in contingent moments.

The categories developed to describe the knowledge required to teach mathematics are underscored by Shulman's (1986, 1987) work. Shulman distinguished two overarching categories of knowledge essential for teaching mathematics:

subject matter knowledge (SMK) and pedagogical content knowledge (PCK); that is, knowledge of what to teach and knowledge of how to teach it (Ernest 1989).

While PCK has been acknowledged in the literature, exactly what it is lacks clarity (Ball et al. 2008). In many instances it is not clear how it is distinguished from other forms of teacher knowledge. It has often been described as the intersection between the subject knowledge and knowledge of teaching and learning (Niess 2005). Shulman (1987) claimed that pedagogical content knowledge comprises specific knowledge that helps teachers to translate and communicate the subject matter to the students they teach. This includes knowledge of instructional strategies and representations, and knowledge of students' conceptions and misconceptions (Shulman 1987). Krauss and colleagues (2008) further clarified PCK as knowledge of mathematical tasks as instructional tools, knowledge of multiple representations and explanations of mathematics problems, and knowledge of students' thinking and explanations. Curricular knowledge encompasses the scope and sequence of teaching programs, and the resources required to implement them (Rowland et al. 2005). Thus, there is agreement that PCK includes multiple components including knowledge of student's conceptions and difficulties, knowledge of instructional strategies, knowledge of curriculum and context, and general pedagogical knowledge (Depaepe et al. 2013).

There has been much debate as to whether pedagogical content knowledge (PCK) and mathematics content knowledge (MCK) are distinct (e.g., Baumert et al. 2010) or overlapping categories (e.g., Ball et al. 2008). However, there has been agreement that strong knowledge of the subject being taught is fundamental to teacher competence (Baumert et al. 2010; Chick et al. 2006; Hill et al. 2005). Quality teachers exhibit MCK prior to and beyond the level they are teaching. Additionally, teachers' MCK together with their confidence in their MCK is strongly related to students' positive gains in their understanding of mathematics and motivation to participate in mathematics in the early years of school (Baumert et al. 2010; Hill et al. 2005). In these settings, teachers who are confident in their subject matter knowledge are more likely to recognize and maximize young children's potential to learn (Hedges and Cullen 2005). Conversely early childhood teachers who are not comfortable with their level of subject matter knowledge tend to rarely include this content in their learning environments or extend their children's thinking in this content area (Anning and Edwards 1999).

PCK and MCK are related constructs. It has been evidenced that having deep mathematics content knowledge does not necessarily result in deep pedagogical content knowledge, but obtaining deep pedagogical knowledge heavily relies on possessing deep mathematics content knowledge (Baumert et al. 2010). In addition, the repertoire of teaching strategies and pool of alternative representations and explanations teachers possess is dependent on the depth of their mathematics content knowledge (Baumert et al. 2010). Teachers who know and use an extensive repertoire of teaching strategies tend to have a deeper understanding of their subject matter knowledge.

PCK and MCK change according to the experiences teachers are involved in during their educational journey. The greatest gains in teachers' subject matter knowledge (SMK) and pedagogical content knowledge (PCK) occur during their pre-service years, and the gains made once teachers are teaching are predominantly for PCK (Kleickmann et al. 2013). There are also relationships between the PCK teachers possess, their experience, and the context in which they teach. Teachers who have taught for more than 10 years seem to have higher PCK than their less experienced counterparts (Lee 2010). In addition teachers with less MCK tend to be situated in low socio-economic status contexts (Hill et al. 2005), and traditional professional development has little impact on the development of teachers' PCK and SMK (Darling-Hammond and Richardson 2009; Kleickmann et al. 2013).

Thus, gaining MCK and associated PCK for teachers whose knowledge of both is low, is a difficult task. Mathematics is based on abstract constructs and teachers have the challenging task of making these constructs accessible for students. Providing students with a 'hands-on approach' to mathematics using appropriate concrete representations is one way of beginning to make mathematics accessible for students in marginalized contexts. This approach moves students away from reading a text that they are unfamiliar with or listening to a teacher deliver a mathematical lesson in a language that is not their first language, to a mode of engagement involving the use of high quality representations.

The Role of Representations in the Teaching and Learning of Mathematics

There are two types of representations namely, *external representations* or *internal representations*, and each has a role to play in developing our understanding of mathematics. Internal representations are mental models or images constructed in the mind. These images include verbal/syntactic, imagistic, formal notational, visual, kinaesthetic and affective (Goldin 2002). External representations are those that are physically embodied or observed such as graphs, number lines, equations, table of values (Goldin and Kaput 1996), and other concrete materials used in the teaching of mathematics. Duval (2006) claims that external representations comprise four registers of mathematics: natural language, figures/diagrams, mono-functional registers of notation systems (symbols), and graphs. He argues that mathematics comprehension results from the coordination or mapping of at least two of these registers. Figure 2.1 provides an example of some of the external representations utilized in an exploration of functional thinking.

Both teachers and students engage with external and internal representations in the teaching and learning process. As students engage with external representations presented by the teacher, they interpret and internalise these representations. Students then communicate their understanding of these representations to others

Function Machine

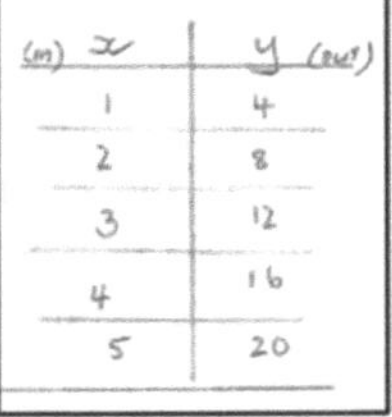

(in) x	y (out)
1	4
2	8
3	12
4	16
5	20

Table

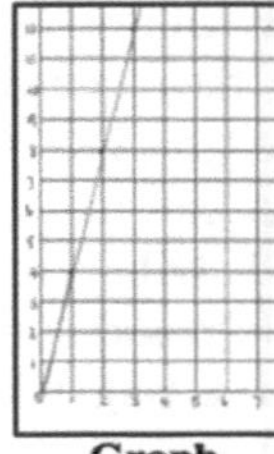

Graph

Fig. 2.1 External representations of functional thinking

through the use of additional external representations. In turn, the teacher internalises the communication from the student to inform their next point of instruction. Thus, each representation provides a new layer of understanding and this cycling continues as students work towards constructing their own understanding of mathematical concepts.

While it has been acknowledged that engaging with multiple external representations provides students with the opportunity to develop a deep understanding of mathematics (Duval 2006; Goldin and Shteingold 2001; Janvier 1987; Tall 1991), there is some debate with regard to how this best occurs. Currently, there are two approaches as to helping students explore multiple external representations for a particular concept. The first introduces the representations simultaneously, so the students engage with a number of different external representations in the one learning experience. It is conjectured that this approach does not provide students with the opportunity to develop a sound understanding of one representation before transferring this to the next representation (Ainsworth et al. 1996). The second approach encourages students to map across representations serving two purposes. It allows students to (a) access a mathematical concept that was not apparent in the previous representation, and (b) deepen their understanding of the concept itself (Duval 2006). A shortfall of the first approach is that it does not provide students with the opportunity to develop a sound understanding of one representation before transferring this to the next representation (Ainsworth et al. 1996). In fact it can lead to confusion rather than development of a deep understanding of a concept. It is the second approach that underpinned the development of the materials utilised in this project.

It is essential for teachers to explicitly map between multiple representations for students to develop deep conceptual understandings of mathematics (Cooper and Warren 2008; Dreyfus 1991; Duval 2006; Halford 1993). As students engage with a new representation they must learn how to use it, and how it links to previous representations. This mapping provides students with the opportunity to transfer their knowledge across different mathematical registers (Duval 2006; Halford 1987). Teachers connect the mathematics for students by linking representations to particular mathematical language. Thus, as students engage with different representations, teachers need to explicitly integrate the mathematical language that links the representations to the concept being taught. This bundling of rich mathematical representations and language provides enriched learning experiences for students.

The Role of Language in Mathematics

In the early years of schooling an oral language approach is strongly related to numeracy development (e.g., Sarama et al. 2012; Warren and Miller 2013). An oral language approach is more than just communicating orally. It entails speaking and listening, and comprehending what is being said, understanding the vocabulary being used, and applying this to mathematical contexts. An oral language approach is not only foundational to literacy (Aldridge 2005), but also underpins the development of mathematical understanding (Krause et al. 2010). Mathematics has its own vocabulary that links specifically to knowledge about concepts and process. It is not enough for students to know definitions of concepts in mathematics; rather they need a deep understanding of the language, concepts and representations. While which words are presented and how they are developed is vitally important (Khisty and Chval 2002), it is not just an emphasis on vocabulary that is required. We cannot disregard other linguistic and cultural elements when helping students construct meaning in the context of mathematics (Celedon-Pattichis et al. 2010).

Mathematical language can be viewed as either a resource or as a political construct (Jorgensen 2011). The former stresses language as supporting thinking, doing, learning and teaching mathematics, whereas the latter refers to language as having the potential to position and transform by placing certain languages and speakers at a disadvantage (Planas and Civil 2013). While we acknowledge the tension between the language required to learn mathematics and the language these students use in and out of the classroom, in this book we are specifically focusing on mathematical language as a resource.

Mathematical language from a linguistic perspective is encapsulated in the literature by the term 'mathematical register' (Halliday 1978). While activities like counting, measuring, and ordering and comparing draw on 'everyday' language, this language in these contexts often serves different functions. Halliday uses the term to describe 'a set of meanings that is appropriate to a particular function of language, together with word and structures that express their meanings' (Halliday 1978, p. 195). This requires teachers to use mathematical language in a certain type of way. From this perspective, learning mathematics involves learning the register, its field, tenor and mode. These refer to the social activity that is occurring (what is being said), the roles and relationships between the participants (beliefs and attitudes), and how the interaction happens (speaking, writing, representing).

The language of instruction impacts on students' access to mathematical concepts. When the language of schooling is Standard Australian English (SAE), the mathematical register used by teachers consists of words that come from two primary sources: (i) everyday English: and (ii) mathematics. The words from everyday English may have the same meaning when used in the mathematics register (e.g. *increase*), may have a different meaning (e.g. *table*), or may have a subtly different nuance (e.g. *between*). Some words sourced from the discipline of mathematics seem to only have meaning in mathematics, including words such as *pronumeral*. In order to be positioned to engage with school mathematics, teaching and assessing using SAE, students need to possess an adequate linguistic repertoire (Meaney et al. 2008).

It is the syntactic, semantic and pragmatic features of the mathematical register that causes most problems for ESL and Indigenous students (Spanos et al. 1988; Niesche 2009). Syntactic features that prove difficult are comparisons of sizes, conditional relationships, and the use of the passive voice (Abedi and Lord 2001; Galligan 1997; Fernandes 2011). The pragmatic features refer to the use of language in particular contexts where culturally specific meanings are required to understand the problem. A prime example of this within the Queensland testing regime was the inclusion of the words *rover pass*, referring to a special ticket that allows all day travel on buses, trains or ferries. Indigenous students from a remote community, unfamiliar with metropolitan contexts, interpreted these words as something about a dog (named Rover) (Baturo et al. 2008).

As Jones et al. (1995) contest, it is not enough to acknowledge just the cultural and linguistic factors that influence the learning of Western mathematics. Such a focus can inappropriately result in a concentration on the use of language rather than the acquisition of mathematical concepts (Howard 1997).

In conclusion, it is conjectured that the main dimensions that impact on teachers' enactment of mathematics teaching in marginalized contexts, are the beliefs and attitudes they bring to these contexts and the mathematical knowledge they possess. The latter includes their knowledge of mathematics content (including its representations and language), and knowledge of how to effectively implement this content in these contexts. These dimensions underpinned the resources we created for use in classrooms. These dimensions also informed the development of the professional learning model we used across the four-year period.

Quality Resources and Activities for Teachers and Students

The principles of equitable teaching also drove the creation of RoleM resources. Equitable teaching is perceived to consist of four dimensions: access, achievement, identity and power (Gutiérrez 2012). Moschkovich (2013) argues that equitable teaching practices for students (a) support mathematical reasoning, conceptual understanding and discourse, and (b) broaden participation for students in learning mathematics. This required ensuring that the resources were: conceptually orientated; open-ended to cater for the differential that exists in students' ability; of high cognitive demand; and culturally appropriate (Boaler and Staples 2008).

Each activity exhibited the following characteristics:

- focused on particular mathematical concepts reflecting high stakes mathematics.
- delineated the specific language and representations required to fully explore the concepts.
- encompassed a degree of flexibility to cater for the learning that was context specific.
- encouraged both direct teaching and group work.
- incorporated a range of representations with a range of hands-on materials.
- allowed for easy differentiation to cater for a range of differences in prior learning and multiple entry points.
- ensured they were culturally appropriate for the different contexts.

Figure 2.2 presents a snapshot of Year 3 learning activities on teaching thirds.

The Indigenous teacher aides that were incorporated in all aspects of the projects played a vital role in ensuring that the activities were culturally appropriate.

Mathematics content knowledge

Purpose of the learning experiences
Often students draw on whole-number understanding to make sense of fractions.... Some students interpret 2/3 as a number between 2 and 3. These misunderstandings occur because they have little experience in partitioning the length model or connecting the length model to the number line.

Making connections
As we move through mathematics the number system expands. ... All whole numbers are fractions.... Locating fractions on a number line helps students make this connection

Mathematical language
Vocabulary
Half, third, smaller, larger, equal parts.....
Open-ended task
Is a quarter smaller than a third? Explain how you know
Word problem
There are 12 pens. If I was given one-third of the pens, how many pens would I have? If I was given one quarter of the pens, how many pens would I have?
NAPLAN type question
Look at the chocolate bars. Which bar shows one-third shaded?

Mathematical pedagogical knowledge

Learning experiences
Consolidating

Engaging

Extending

Independent learning

Each includes instructions to consider using, points of discussion and questions that can be asked.

Representations
Paper sheets (circles and rectangles)
Strings
Streamers
Number lines
Digital (Apps)

Teachable moments
... hands on folding activities help students 'see' how fractions are related to each other. After students have folded the string into thirds, mark one third, two thirds. Have them to fold the string in halves.... Ask: *Is two thirds closer to a half or one?*
Mathematical recording
..... Accompany this recording with appropriate language (e.g. four thirds is four out of three equal parts)

Fig. 2.2 Year 3 RoleM learning activities: *Exploring thirds*

Different representations and examples were provided by them that guaranteed the activities closely aligned with the culture and the environment in which they were situated.

In brief, the materials allowed teachers to differentiate the learning by assigning students to different types of activities (e.g., engaging, consolidating, extending) according to their level of understanding. The different types of representations and materials used within the activities also allowed teachers to cater for a wide range of learning styles (e.g. kinesthetic, auditory, visual). The activities encouraged students to explore mathematical concepts through a range of representations (e.g., number lines, number tracks, ten frames). Finally, the activities helped teachers to become aware of encouraging students to be active participators in classroom talk about mathematics (e.g., vocabulary, open-ended tasks, word problems).

Additionally, multiple representations were used to teach each concept in mathematics. As a new representation was introduced in a new learning activity, it was mapped back onto prior representations of that concept. Some representations (e.g., 10 frames) were used across all year levels to develop a deeper conceptual understanding of the multiplicative nature of our number system.

Professional Learning

One of the greatest influences on students' learning is what teachers do in the classroom. Up to 30 % of the variation in students' performance can be attributed to their teachers (Hattie 2003). Thus, improving students' performance in marginalized contexts is closely related to improving the quality of their teachers' teaching. While the relationship between teachers' mathematics content knowledge (MCK) and their teaching performance is complex, there is general consensus that good content knowledge facilitates good teaching and students' progress (Rowland et al. 2000).

Given that mentors in these contexts are few and far between, the strategy that is seen as imperative for these schools to adopt is high quality professional learning for their teachers (Desimone 2009; Hattie 2009). But efforts to bring about change via traditional professional development sessions are insufficient. Professional learning happens over a long time and is a contextualized holistic experience (Darling-Hammond et al. 2009; Vygotsky 1978). We contest that quality professional learning is underscored by theories of learning, and is thereby dependent on the interactions that occur between the learner, the context, and what is learned (Gravani 2007; Goos and Makar 2007; Jarvis and Parker 2005). Additionally, the strength of the support offered by 'experts' from within the school and wider community (Askew 2008) can make a difference to the quality of teaching. But these 'experts' need to have strong mathematical backgrounds and deep understanding of how to effectively teach mathematics.

Supporting the Professional Learning of Teachers in These Contexts

The particular theories that drove the development of this project were socio-cultural perspectives of learning (Vygotsky 1978) and communities of practice (Lave and Wenger 1991). The significance of the Vygostkian perspective is that it extinguishes the traditional boundaries between individual effort and social interactions as the individual comes to 'know'. Learning is situated in a community of practice, a community that involves 'ways of doing things that are shared to some significant extent among members' (Lave and Wenger 1991). In this context, learning as an acquisition of knowledge is situated in social relationships, and involves a process of social participation and enculturation into the wider practices of society (Borko 2004). To master new knowledge and skills newcomers are required to move towards full participation in the socio-cultural practices of the community.

Integral to a socio-cultural perspective is the notion of the Zone of Proximal Development (ZPD) (Vygotsky 1978). ZPD is defined as an individual's potential capacity for development through the assistance of a more knowing person (Vygotsky 1978). The significance of ZPD is that it determines the lower and upper bounds of the zone within which professional development (PD) instruction and teacher learning should be directed. In the lower bounds, formal PD sessions provide important information that teachers need to know about mathematical content, changes in the curriculum, innovative teaching strategies, and using resources effectively. However, instruction is only efficacious when it goes beyond the notion of simply assisting a person to acquire a particular set of skills or knowledge. Such instruction enables learners to extend themselves through active engagement, exploration and investigation of teaching and learning concepts and activities. In the upper bounds of the ZPD, the 'more knowing person', or 'expert', provides support for teachers through mentoring and scaffolding as these teachers are guided towards competent and accomplished practices (Brockbank and McGill 2006). A purported result is that the learner is better placed to independently implement innovative pedagogical practices across all curriculum areas after the 'expert' has withdrawn.

The depth and scope of learning is also influenced by the nature and quality of a teacher's reflection (Phillips 2008; Wells 1999). Thus, when extensive teacher reflection is combined with action, students' experiences are transformed into learning (Schon 1983). Teacher reflection serves both as an instrumental and a critical function (van Manen 1977). The former encourages teachers to reflect on teaching and learning problems that arise in their classrooms, and formulate practical plans that may solve the problem. Reflection, as a critical function, provides cognitive and affective insights that can challenge assumptions teachers hold about things, such as: the nature of teaching and themselves as teacher; and their students' ability as learners in mathematics (van Manen 1977). As Dewey stated, genuine thinking only occurs 'when there is a tendency to doubt' (as cited in Garrison 2006

p. 3). Therefore, effective teacher learning requires multiple cyclical movements among systems of influence and teachers' worlds (Opfer and Pedder 2011). With ongoing support, teachers and 'experts' become co-constructors of knowledge moving within and beyond each other's ZPD.

The RoleM Professional Learning Model

The RoleM professional learning model is a socio-constructivist model based on the theories of Vygotsky (1978). Six principles drawn from the literature also underpin the model:

1. Meaningful learning is a slow and uncertain process for teachers, requiring time and teachers' reflection (e.g., Boyle et al. 2005).
2. Teachers' professional learning is more evident when continuing professional development (PD) includes a focus on teachers' knowledge of the subject (e.g., Borko 2004) and classroom practicalities (e.g., Porter et al. 2000).
3. PD emphasizing general teachers' knowledge and teaching competencies known to improve student learning and requires teachers to reconsider their current practices (e.g., Desimone 2009; Timperley 2008).
4. Professional development is more meaningful to teachers when it is situated within the context of their workplace (Webster-Wright 2009).
5. The most significant changes in teacher beliefs and attitudes occur when teachers have multiple opportunities to absorb new information, put it into practices and observe improved student learning outcomes (e.g., Guskey 2010; Darling-Hammond and Richardson 2009).
6. Resourcing has an impact on a teacher's capacity to effectively teach mathematics (e.g., Clements 2004).

We argue that professional development days are components of professional learning. For effective professional learning to occur teachers, together with experts, need to trial ideas in their classroom contexts and reflect on the student learning that has occurred.

The RoleM professional learning model (RPL) is cyclical in nature and involves teachers in self-reflection as they trial approaches and resources in their classrooms to improve the quality of their teaching practice. It is based on the view that teachers have the ability to improve their practice by trialing 'proven' effective learning experiences, and through continuous cycles of on-the-job reflections and discussions with experts from the field (Castle and Aichele 1994; Clarke and Hollingworth 2002). Our model builds on Guskey's (1988) three domains of practice, consequence and personal. The domain of practices entails professional experimentation with the context of the classroom. The domain of consequence refers to salient outcomes such as improved student learning. The personal domain encompasses ones knowledge, beliefs and attitudes. The model also reflects Clarke and Hollingworth's (2002) notion of effective professional learning is cyclical and

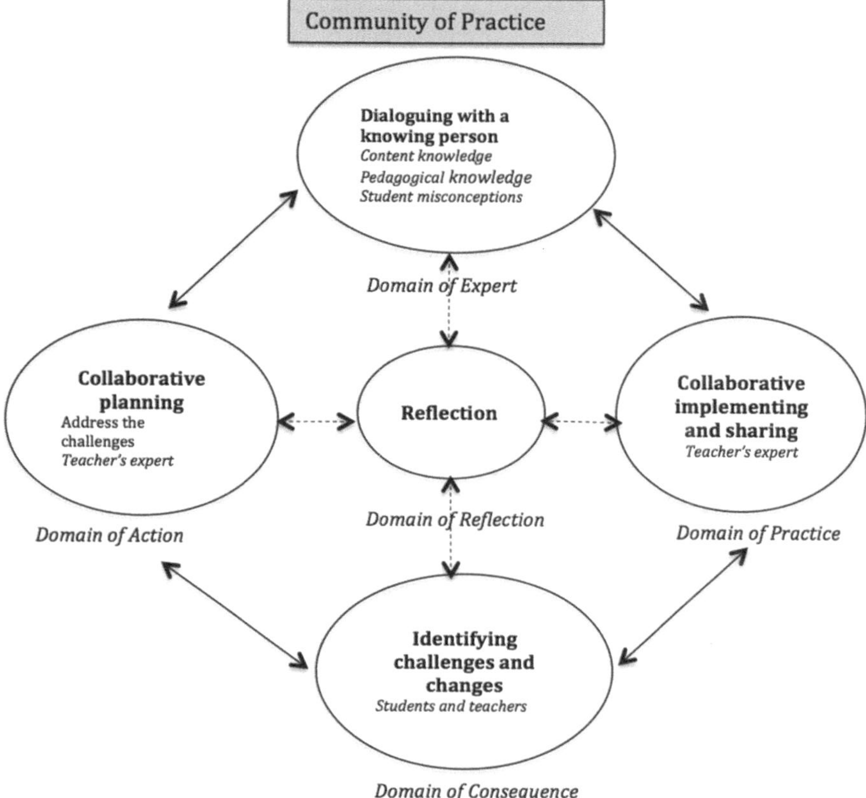

Fig. 2.3 The RoleM professional learning model

the domains are inter-connected. Figure 2.3 presents the key components together with the key focuses, of the professional learning model.

The contribution that this model makes to literature pertaining to effective professional learning is the inclusion of the domain of expert, domain of reflection and domain of action (practice).

Implementation of the Model Within These Disadvantaged Contexts

Teachers and teacher assistants (TAs) attended professional development days three times a year. The professional development days focused on a specific year level for each year (e.g., 2010 Foundation Year, 2011 Year 1). At each professional development day teachers and TAs interacted with 10 to 12 RoleM mathematical learning experiences. The notion of active learning underpinned the professional learning that occurred over the PD day. Teachers 'acted out' the activities and discussed the mathematics content knowledge, pedagogical knowledge, mathematical language and representations that underpinned each. They also reflected on possible student misconceptions and how these could be addressed. At these PD days, teachers also received the RoleM resources ready for implementation in their classrooms.

Approximately three weeks after each PD day, experts conducted follow up visits (FV) in each teacher's classroom. Before arriving at the school, teachers were contacted and asked what mathematical content they would like the expert to focus on. Each follow up day was different, as the format of each was teacher-driven. At these visits experts and teachers worked collaboratively to address any issues identified by the teacher. For example, experts may have: modeled lessons: observed and provided feedback to teachers; worked with TA's; and/or worked with individual students to help diagnose their misconceptions or extend their learning. Finally, teachers and experts worked together to plan the next phase in students' learning, and the cycle started again. These FV also occurred three times a year.

The literature review delineated a series of sub-questions that serve to further illuminate the overarching research question that drove the project, which was:

What is the interplay between effective teaching and effective learning in these contexts?

The sub-questions were:

How do teachers' dispositions towards teaching mathematics influence their classroom practice in these contexts?

How does teachers' mathematics content knowledge and pedagogical knowledge influence their classroom practices in these contexts?

How does professional learning best support the development of this interplay?

References

Abedi, J., & Lord, C. (2001). The language factor in mathematics tests. *Applied Measurement in Education, 14*(3), 219–234.

Ainsworth, S. E., Wood, D. J., & Bibby, P. A. (1996). Co-ordinating multiple representations in computer based learning environments. In P. Brna, A. Paiva, & J. Self (Eds.), *Proceedings of the European conference of artificial intelligence in education* (pp. 336–342). Ediçoes Colibri, Lisbon: ECAI.

Aldridge, J. (2005). The importance of oral language. *Childhood Education, 81*(3), 177–180.

Ambrose, R. (2004). Initiating change in prospective elementary school teachers' orientations to mathematics teaching by building on beliefs. *Journal of Mathematics Teacher Education, 7*, 91–119.

Anning, A., & Edwards, A. (1999). *Promoting Children's Learning from Birth to Five: Developing the new early years professional*. Buckingham: Open University Press.

Ashton, P., & Webb, R. (1986). *Making a difference: Teachers' sense of efficacy and student achievement*. New York: Longman.

Askew, M. (2008). Mathematical discipline knowledge requirements for prospective primary teachers, and the structure and teaching approaches of programs designed to develop that knowledge. In K. Krainer & T. Wood (Eds.), *The international handbook of mathematics teacher education* (Vol. 1, pp. 13–36). Rotterdam: Sense Publishers.

Ball, D. L., Thames, M. H., & Phelps, G. (2008). Content knowledge for teaching what makes it special? *Journal Of Teacher Education, 59*(5), 389–407.

Bandura, A. (1991). Social cognitive theory of self-regulation. *Organizational Behavior and Human Decision Processes, 50*(2), 248–287.

Bandura, A. (1992). Self-efficacy mechanism in psychobiologic functioning. In R. Schwarzer (Ed.), *Self-efficacy: Thought control of action* (pp. 355–394). Washington, DC: Hemisphere.

Baturo, A., Cooper, T., Michaelson, M., & Stevenson, J. (2008). Using national numeracy testing to benefit Indigenous students: Case studies of teachers taking back control of outcomes. In M. Goos, R. Brown, & K. Makar (Eds.), *Navigating currents and charting directions (Proceedings of the 31st Annual Conference of the Mathematics Education Research Group of Australasia)* (pp. 59–66). Brisbane, Qld: MERGA.

Baumert, J., Kunter, M., Blum, W., Brunner, M., Voss, T., Jordan, A., et al. (2010). Teachers' mathematical knowledge, cognitive activation in the classroom, and student progress. *American Educational Research Journal, 47*(1), 133–180.

Beswick, K., Watson, J., & Brown, N. (2006). Teachers' confidence and beliefs and their students' attitudes to mathematics. In P. Grootenboer, R. Zevenbergen, & M. Chinnappan (Eds.), *Identities, cultures and learning spaces (Proceedings of the 29th annual conference of the mathematics education research group of Australasia* (Vol. 1, pp. 68–75). Adelaide: Mathematics Education Research Group of Australasia.

Boaler, J., & Staples, M. (2008). Creating mathematical futures through an equitable teaching approach: The case of railside school. *Teachers' College Record, 110*(3), 608–645.

Borko, H. (2004). Professional development and teacher learning: Mapping the terrain. *Educational Researcher, 33*, 3–15.

Boyle, B., Lamprianou, I., & Boyle, T. (2005). A longitudinal study of teacher change: What makes professional development effective? Report of the second year of the study. *School Effectiveness and School Improvement, 16*(1), 1–27.

Brockbank, A., & McGill, I. (2006). *Facilitating reflective learning through mentoring and coaching*. Philadelphia: Kogan Page.

Castle, K., & Aichele, D. B. (1994). Professional development and teacher autonomy. In D. B. Aichele & A. F. Coxford (Eds.), *Professional development for teachers of mathematics, 1994 yearbook*. National Council of Teachers of Mathematics: Reston Virginia.

Celedon-Pattichis, S., Musanti, S., & Marshall, M. (2010). *Bilingual elementary teachers' reflections on using students' native language and culture to teach mathematics*. Retrieved from www.researchgate.net/publication/266061798.

Chick, H. L., Baker, M., Pham, T., & Cheng, H. (2006). Aspects of teachers' pedagogical content knowledge for decimals. In J. Novotná, H. Moraová, M. Krátká, & N. Stehlíková (Eds.), *Proceedings of the 30th annual conference of the International Group for the Psychology of Mathematics Education* (Vol. 2, pp. 297–304). Prague: PME.

Chick, H., & Beswick, K. (2013). Educating Boris: An examination of pedagogical content knowledge for mathematics teacher educators. In V. Steinle, L. Ball, & C. Bardini (Eds.), *Mathematics education: Yesterday, today and tomorrow (Proceedings of the 36th annual conference of the Mathematics Education Research Group of Australasia)* (pp. 170–177). Melbourne, Vic: MERGA.

Clarke, D. M., Roche, A., & Downton, A. (2009). Assessing teacher pedagogical content knowledge: Challenges and insights. In M. Tzekaki, M. Kaldrimidou, & H. Sakonidis (Eds.), *In Search of Theories in Mathematics Education, Proceedings of the 33rd conference of the International Group for the Psychology of Mathematics Education* (pp. 362–369). Thessaloniki, Greece: PME.

Clarke, D., & Hollingsworth, H. (2002). Elaborating a model of teacher professional growth. *Teaching and Teacher Education, 18*(8), 947–967.

Clements, D. H. (2004). Major themes and recommendations. In D. H. Clements & J. Sarama (Eds.), *Engaging young children in mathematics: Standards for early childhood mathematics education* (pp. 7–72). Mahwah: Erlbaum.

Cooper, T. J., & Warren, E. (2008). The effect of different representations on Years 3 to 5 students' ability to generalise. *ZDM, 40*(1), 23–37.

Darling-Hammond, L., & Richardson, N. (2009). Teacher learning: What matters? *Educational Leadership, 66*(5), 46–53.

Darling-Hammond, L., Wei, R. C., Andree, A., Richardson, N., & Orphanos, S. (2009). *Professional learning in the learning profession*. Washington, DC: National Staff Development Council.

Depaepe, F., Verschaffel, L., & Kelchtermans, G. (2013). Pedagogical content knowledge: A systematic review of the way in which the concept has pervaded mathematics educational research. *Teaching and Teacher Education, 34*, 12–25.

Desimone, L. M. (2009). Improving impact studies of teachers' professional development: Toward better. *Educational Researcher, 38*(3), 181–199.

DeBellis, V. A., & Goldin, G. A. (2006). Affect and meta-affect in mathematical problem solving: A representational perspective. *Educational Studies in Mathematics, 63*(2), 131–147.

Dreyfus, T. (1991). Advanced mathematical thinking processes. In D. Tall (Ed.), *Advanced mathematical thinking* (pp. 25–41). Springer: Netherlands.

Duval, R. (2006). A cognitive analysis of problems of comprehension in a learning of mathematics. *Educational studies in mathematics, 61*(1–2), 103–131.

Ernest, P. (1988). The attitudes and practices of student teachers of primary school mathematics. In A. Borbas (Ed.), *Proceedings of 12th International Psychology of Mathematics Education Conference* (pp. 288–295). Veszprem, Hungary: PME.

Ernest, P. (1989). The knowledge, beliefs and attitudes of the mathematics teacher: a model. *Journal of Education for Teaching, 15*, 13–33.

Fernandes, A. (2011). Mathematics pre-service teachers learning about English language learners through task-based interviews. In J. Clark, B. Kissane, J. Mousley, T. Spencer, & S. Thornton (Eds.), *Mathematics: Traditions and [new] practices (Proceedings of the 34th Annual Conference of the Mathematics Education Research Group of Australasia and the 23rd biennial conference of the Australian Association of Mathematics Teachers* (pp. 235–243). Adelaide: AAMT & MERGA.

Galligan, L. (1997). Relational word problems: a cross-cultural comparison. In A. Begg (Ed.), *People in mathematics education (Proceedings of the 20th Annual Conference of the Mathematics Education Research Group of Australasia)* (pp. 177–183). Rotorua: MERGA.

Garrison, J. (2006). The "permanent deposit" of Hegelian thought in Dewey's theory of inquiry. *Educational Theory, 56*(1), 1–37.

Goldin, G. A. (2002). Representation in mathematical learning and problem solving. In L. D. English (Ed.), *Handbook of international research in mathematics education* (pp. 197–218). Mahwah, New Jersey: Lawrence Erlbaum Associates.

Goldin, G., & Kaput, J. (1996). A joint perspective on the idea of representations in learning and doing mathematics. In S. P. Leslie & N. Pearla (Eds.), *Theories of mathematical learning*. Mahwah NJ: Lawrence Erlbaum Associates.

Goldin, G., & Shteingold, N. (2001). Systems of representations and the development of mathematical concepts. In A. A. Cuoco & F. R. Curcio (Eds.), *The roles of representation in school mathematics* (pp. 1–23). Reston, VA: NCTM.

Goos, M., & Makar, K. (2007). Designing professional development to support teachers' learning in complex environments. *Mathematics Teacher and Development, 8*, 23–47.

Gravani, M. (2007). Unveiling professional learning: Shifting from the delivery of courses to an understanding of the processes. *Teaching and Teacher Education, 23*, 688–704. Gresham,

Grootenboer, P., Lomas, G., & Ingram, N. (2008). The affective domain and mathematics education. In H. Forgasz, A. Barkatsas, A. Bishop, B. Clarke, S. Keast, W. T. Seah, & P. Sullivan (Eds.), *Research in mathematics education in Australasia 2004–2007* (pp. 255–269). Amsterdam: Sense.

Gutiérrez, R. (2012). Context matters: How should we conceptualize equity in mathematics education? In B. Herbel-Eisenmann, J. Choppin, D. Wagner, & D. Pimm (Eds.), *Equity in discourse for mathematics education: Theories, practices, and policies*. New York, NY: Springer.

Guskey, T. R. (1988). Teacher efficacy, self-concept, and attitudes toward the implementation of instructional innovation. *Teaching and Teacher Education, 4*, 63–69.

Guskey, T. R. (2010). Professional development and teacher change. *Teachers and Teaching, 8*(3), 381–391.

Hannula, M. (2002). Attitude towards mathematics: Emotions, expectations and values. *Educational Studies in Mathematics, 49*, 25–46.

Halford, G. (1993). *Children's understanding: The development of mental models*. Hillsdale, NJ: Lawrence Erlbaum Associates.

Halford, G. (1987). A structure-mapping approach to cognitive development. *International Journal of Psychology, 22*, 609–642.

Halliday, M. A. K. (1978). *Language as social semiotic*. London: Arnold.

Hattie, J. (2003, October). *Teachers make a difference: What is the research evidence?* Keynote presentation at the Building Teacher Quality: The ACER Annual Conference, Melbourne, Australia. Retrieved from http://www.acer.edu.au/documents/TeachersMakeaDifferenceHattie.doc.

Hattie, J. (2009). *Visible learning: A synthesis of 800+ meta-analyses on achievement*. Abingdon: Routledge.

Hedges, H., & Cullen, J. (2005). Subject knowledge in early childhood curriculum and pedagogy: Beliefs and practices. *Contemporary Issues in Early Childhood, 6*(1), 66–79.

Hill, H. C., Rowan, B., & Ball, D. L. (2005). Effects of teachers' mathematical knowledge for teaching on student achievement. *American Educational Research Journal, 42*(2), 371–406.

Howard, P. (1997, November). *Aboriginal voices in our schools*. Paper presented at the Annual Conference of the Australian Association for Research in Education, Brisbane, Australia.

Janvier, C. E. (1987). *Problems of representation in the teaching and learning of mathematics*. Hillside, NJ: Lawrence Erlbaum Associates.

Jarvis, P., & Parker, S. (Eds.). (2005). *Human learning: A holistic approach*. New York: Routledge.

Jones, K., Kershaw, L., & Sparrow, L. (1995). *Aboriginal children learning mathematics*. Perth, Western Australia: MASTEC, Edith Cowan University.

Jorgensen, R. (2011). Language, culture and learning mathematics: A Bourdieuian analysis of Indigenous learning. In C. Wyatt-Smith, J. Elkins, & S. Gunn (Eds.), *Multiple perspectives on difficulties in learning literacy and numeracy* (pp. 315–329). Dordrecht, The Netherlands: Springer.

Jorgensen, R., Grootenboer, P., Niesche, R., & Lerman, S. (2010). Challenges for teacher education: The mismatch between beliefs and practice in remote Indigenous contexts. *Asia-Pacific Journal of Teacher Education, 38*(2), 161–175.

Kagan, D. M. (1992). Implications of research on teacher belief. *Educational Psychologist, 27*, 65–90.

Khisty, L., & Chval, K. (2002). Pedagogic discourse and equity in mathematics: When teachers' talk matters. *Mathematics Education Research Journal, 14*(3), 4.18.

Kleickmann, T., Richter, D., Kunter, M., Elsner, J., Besser, M., Krauss, S., & Baumert, J. (2013). Teachers' content knowledge and pedagogical content knowledge the role of structural differences in teacher education. *Journal of Teacher Education, 64*(1), 90–106.

Krause, K.-L., Bochner, S., Duchesne, S., & McMaugh, A. (2010). Educational psychology for learning and teaching (3rd ed.). Southbank, Victoria: Cengage Learning Australia Ltd.

Krauss, S., Brunner, M., Kunter, M., Baumert, J., Blum, W., Neubrand, M., & Jordan, A. (2008). Pedagogical content knowledge and content knowledge of secondary mathematics teachers. *Journal of Educational Psychology, 100*(3), 716–725.

Lave, J., & Wenger, E. (1991). *Situated learning: Legitimate peripheral participation*. Cambridge: Cambridge University Press.

Lee, J. (2010). Exploring kindergarten teachers' pedagogical content knowledge of mathematics. *International Journal of Early Childhood, 42*(1), 27–41.

Ma, X., & Kishor, N. (1997). Assessing the relationship between attitude toward mathematics and achievement in mathematics: A meta-analysis. *Journal for Research in Mathematics Education, 28*(1), 26–47.

Matthews, C., Watego, L., Cooper, T. J., & Baturo, A. R. (2005). Does mathematics education in Australia devalue Indigenous culture? Indigenous perspectives and non-Indigenous reflections. In P. Clarkson, A. Downtown, D. Gronn, M. Horne, A. McDonough, R. Pierce, & A. Roche (Eds.), *Proceedings of the 28th Annual Conference of the Mathematics Education Research Group of Australasia* (pp. 513–520). Melbourne, Vic.: University of Melbourne.

McLeod, D. B. (1992). Research on affect in mathematics education: A reconceptualization. In D. A. Grouws (Ed.), *Handbook of research on mathematics learning and teaching* (pp. 575–596). New York: MacMillan.

Meaney, T., Fairhill, U., & Trinick, T. (2008). The role of language in ethnomathematics. *The Journal of Mathematics and Culture, 3*(1), 52–65.

Moschkovich, J. (2013). Principles and guidelines for equitable mathematics teaching practices and materials for english language learners. *Journal of Urban Mathematics Education, 6*(1), 45–57.

Niesche, R. (2009). The use of home language in the mathematics classroom. In R. Hunter, B. Bicknell, & T. Burgess (Eds.), *Proceedings of the 32nd Annual Conference of the Mathematics Education Research Group of Australasia* (pp. 704–707). Palmerston North, NZ: MERGA.

Niess, M. (2005). Preparing teachers to teach science and mathematics with technology: Developing a technology pedagogical content knowledge. *Teacher and Teacher Education, 21*, 509–523.

Opfer, V. D., & Pedder, D. (2011). Conceptualizing teacher professional learning. *Review of Educational Research, 81*(3), 376–407.

Phillips, P. (2008). Professional development as a critical component of continuing teacher quality. *Australian Journal of Teacher Education, 33*(1), 1–9.

Philipp, R. A. (2007). Mathematics teachers' beliefs and affect. In F. K. Lester Jr (Ed.), *Second handbook of research in mathematics teaching and learning* (pp. 257–315). Reston, VA: National Council of Teachers of Mathematics.

Planas, N., & Civil, M. (2013). Language-as-resource and language-as-political: Tensions in the bilingual mathematics classroom. *Mathematics Education Research Journal, 25*(3), 361–378.

Porter, A., Garet, M., Desimone, L., Birman, B., & Yoon, K. S. (2000). *Does professional development change teaching practice? Results from a three-year study*. Washington: U.S. Department of Education.

Rohrkemper, M. (1986). The functions of inner speech in elementary school students' problem-solving behavior. *American Educational Research Journal, 23*(2), 303–313.

Rowland, T., Martyn, S., Barber, P., & Heal, C. (2000). Primary teacher trainees' mathematics subject knowledge and classroom performance. *Research in mathematics education, 2*(1), 3–18.

Rowland, T., Huckstep, P., & Thwaites, A. (2005). Elementary teachers' mathematics subject knowledge: The knowledge quartet and the case of Naomi. *Journal of Mathematics Teacher Education, 8*(3), 255–281.

Sarama, J., Lange, A. A., Clements, D. H., & Wolfe, C. B. (2012). The impacts of an early mathematics curriculum on oral language and literacy. *Early Childhood Research Quarterly, 27*(3), 489–502.

Schon, D. A. (1983). *The reflective practitioner: How professionals think in action*. New York: Basic Books.

Schuck, S., & Grootenboer, P. (2004). Affective issues in mathematics education. In B. Perry, C. Diezmann, & G. Anthony (Eds.), *Review of mathematics education in Australasia 2000–2003* (pp. 53–74). Sydney: MERGA.

Shulman, L. S. (1986). Those who understand: Knowledge growth in teaching. *Educational Researcher, 15*(2), 4–14.

Shulman, L. S. (1987). Knowledge and teaching: Foundations of the new reform. *Harvard educational review, 57*(1), 1–23.

Spanos, G., Rhodes, N. C., Dale, T. C., & Crandall, J. (1988). Linguistic features of mathematical problem solving: Insights and applications. In R. R. Cocking & J. P. Mestre (Eds.), *Linguistic and cultural differences on learning mathematics* (pp. 221–240). Hillsdale: Erlbaum.

Staub, F., & Stern, E. (2002). The nature of teacher's pedagogical content beliefs matters for students' achievement gains: Quasi-experimental evidence from elementary mathematics. *Journal of Educational Psychology, 94*(2), 344–355.

Stein, M. K., & Lane, S. (1996). Instructional tasks and the development of student capacity to think and reason: An analysis of the relationship between teaching and learning in a reform mathematics project. *Educational Research and Evaluation, 2*(1), 50–80.

Stipek, D., Givvin, K., Salmon, J., & MacGyvers, V. (2001). Teachers' beliefs and practices related to mathematics instruction. *Teaching and Teacher Education., 17*(2), 213–226.

Sullivan, P., Clarke, D., Clarke, B., & O'Shea, H. (2009). Exploring the relationship between tasks, teacher actions, and student learning. In M., Tzekaki, M., Kaldrimidou, H., Sakonidis, H (Eds.), *Search of Theories in Mathematics Education, Proceedings of the 33rd Conference of the International Group for the Psychology of Mathematics Education* (pp. 185–195). Thessaloniki, Greece: PME.

Tall, D. (Ed.). (1991). *Advanced mathematical thinking*. Netherlands: Kluwer Academic Publishers.

Timperley, H. (2008). *Teacher professional learning and development*. Brussels: International Academy of Education.

Van Manen, M. (1977). Linking ways of knowing with ways of being practical. *Curriculum Inquiry, 6*(3), 205–228.

Vygotsky, L. S. (1978). *Mind in society: The development of higher psychological processes*. Cambridge, MA: Harvard University Press.

Warren, E., & Miller, J. (2013). Young Australian Indigenous students' effective engagement in mathematics: The role of language, patterns, and structure. *Mathematics Education Research Journal, 25*(1), 151–171.

Webster-Wright, A. (2009). Reframing professional development through understanding authentic professional learning. *Review of Educational Research, 79*(2), 702–740.

Wells, G. (1999). *Dialogic inquiry: Toward a sociocultural practice and theory of education*. New York: Cambridge University Press.

Chapter 3
Mathematics and Marginality

Abstract Mathematics is a content area that many elementary teachers struggle with in the classroom. For many teachers their self-efficacy with regard to their own ability to do mathematics is low (Hoy 2000). This struggle is even more pronounced in the early year classrooms in the elementary school, with a consequence that many childhood teachers focus their teaching on literacy at the expense of mathematics (Frigo et al. 2004; Jurdak 2009). Teaching in marginalized contexts only adds to this problem, especially for teachers who are new to the profession and hold feelings of not being able to cope. This chapter begins by sharing the journey of teachers working in these contexts as they progress through the RoleM professional learning. It also delineates the strategies used in RoleM that helped them become quality teachers.

One dimension that appears to make a significant difference to students' learning: is the knowledge that teachers have about teaching mathematics. Knowledgeable teachers can have a substantive effect on all students' learning including students from marginalized contexts, and this impact is consistent across all levels of schooling including the early years (Hill et al. 2005). The knowledge that teachers have about mathematics impacts on their teaching of mathematics. Teachers whose knowledge is limited tend to place greater emphasis on rote learning and the teaching of facts, and rely more heavily on workbooks and worksheets (Aubrey 2003). Less knowledgeable and qualified teachers appear to be situated more prevalently in marginalized contexts (Darling-Hammond and Post 2000; Hill et al. 2005).

Marginalized contexts experience difficulties in attracting and retaining teachers (OECD 2012; UNESCO 2014), and the teachers that they do attract often receive little quality support in the way of mentoring or induction (Darling-Hammond and Post 2000). This could be due to the lack of expertise within the context itself, as within these contexts good teachers are scarce (Lupton 2004). Also, most teachers working in these contexts lack the preparation to take on mentoring roles (Ayalon 2011). In addition, most induction programs tend to have a curriculum focus

E. Warren and J. Miller, *Mathematics at the Margins*,
SpringerBriefs in Education, DOI 10.1007/978-981-10-0703-3_3

(Feiman-Nemser 2001) rather than a context focus. Thus for many teachers in these contexts their experience becomes one of survival on the run. They quickly adopt practices that they experienced at school or practices that seem to work in the moment rather than selecting 'best' practice for their students, and these survival tactics quickly become entrenched (Feiman-Nemser 2001). The support that these teachers receive is, at best, sporadic.

Participating Teachers

Across the four-year period the sample comprised 154 teachers with over half located in the metropolitan schools. Teachers were categorized as being inexperienced (taught 0–2 years), experienced (taught 3–9 years), or very experienced (taught over 10 years). Table 3.1 presents the frequency of teachers in each category in each geographical location.

Fifty-five percent of the teachers were inexperienced with 46 % of these teachers teaching in remote and very remote geographical locations. A third of these inexperienced teachers were new graduates. The distribution of teachers across the four year levels was relatively consistent, with each year level (Foundation to Year 4) comprising approximately 55 % inexperienced teachers, 23 % experienced teachers, and 22 % very experienced teachers.

While there were 154 teachers who participated in RoleM professional learning, only 98 teachers were interviewed. Over the four-year period the focus of RoleM professional learning moved from Foundation teachers to Year 3 teachers. Each year new resources were developed and distributed to the appropriate year level teachers. For example, in the first year Foundation materials were developed and distributed, in the second year Year 1 materials were developed and distributed, and so on. Teachers who had already participated in RoleM at these schools continued to use and adapt the RoleM materials provided to them in their year of participation. New teachers in these year levels were provided with the RoleM materials and relied on their peers to help them implement the materials in their classroom. Only the teachers who were actively participating in the RoleM professional learning were interviewed. Table 3.2 presents the number of teachers interviewed over the four-year period.

Thus over the four year period 73 teachers were interviewed three times in their year of participation. The findings presented in this chapter emerged from the

Table 3.1 Frequencies of teachers experience in relation to geographical location

Location	Inexperienced	Experienced	Very experienced	Total
Metropolitan	45	24	20	89
Remote	26	8	11	45
Very remote	13	3	4	20
	84	35	35	154

Table 3.2 Frequency of teachers interviewed: 2010–2013 (n = 98)

Year	Three interviews	Two interviews	One interview	Total
2010	21	3	0	24
2011	20	4	2	26
2012	15	7	2	24
2013	17	5	2	24
Total	73	19	6	98

Table 3.3 Frequency of teachers interviewed 3 times by experience and location (n = 73)

Location	Inexperienced	Experienced	Very experienced	Total
Metropolitan	22	10	10	42
Remote	8	8	3	19
Very remote	8	1	3	12
	38	19	16	73

analysis of the three interviews conducted with these 73 teachers. The 73 teachers were spread across locations and had varying levels of experience. Table 3.3 presents the spread of teachers according to their location and experience.

Over half the teachers interviewed were from metropolitan schools, and over half the teachers were inexperienced, with the highest percentage of inexperienced teachers working in schools situated in very remote locations.

Data Collection and Analysis

Seventy-three teachers participated in three semi-structured interviews throughout each calendar year. A semi-structured interview is often described as 'a conversation with a purpose' (Smith et al. 2009 p. 57). It provides a space for participants to tell their own story and allows for the collection of large quantities of rich data. Three broad themes were embedded in each teacher interview: mathematical knowledge and understanding; perceptions of student learning and abilities; and current pedagogical practices. These interviews were conducted three weeks after the professional learning day had occurred (see Chap. 2). Interviews were of 30 min duration and conducted by telephone at a convenient time for the participants. All interviews were audio-recorded for later transcription.

The data were analyzed using a grounded methodological approach (Strauss and Corbin 1998). The transcribed texts were examined in an attempt to identify teachers' self-reported beliefs and practices. This process used an *open-coding* approach to break the data into distinct units of meaning. The application of a *constant comparative* method allowed patterns and themes to emerge (Strauss and Corbin 1990). This approach incorporates a progression from merely describing what is happening to explaining the relationship between and across incidents. Researchers conducted independent member-checks on the emergent themes and subthemes, and compared the data across the interviews. Where disagreement existed, researchers returned to the raw data gathering excerpts of their stances until a consensus was reached. Each theme is represented by a series of quotes taken from within the transcripts. For the purpose of this book, in the first instance these themes and quotes are used to generate narratives representing the different types of teachers (beginning and experienced) in different contexts (very remote, remote and metropolitan). These narratives represent two junctures of their journey, the beginning and the end. These two junctures were chosen as they evidence the extent to which teachers changed after one years participation with the RoleM project.

Results

Beginning of Their RoleM Professional Learning Journey

To gauge the issues that teachers experienced at the commencement of RoleM professional learning, an analysis was conducted on the first interview of the 21 teachers who were interviewed three times in 2010 (the first year of RoleM). This sample comprised 13 teachers from the metropolitan schools (4 of whom were very experienced), 4 teachers from the remote schools, and 4 teachers from the very remote schools. To ascertain if the issues they faced differed from location to location, the interviews were grouped according to the teachers' locations and their prior experience. As all 4 teachers from the very remote schools were all inexperienced and 3 of the teachers from the remote schools were experienced, it was decided to compare and contrast their responses to groups of like teachers from the metropolitan schools. Thus, 16 teachers' transcripts informed the development of these stories. The next section presents a narrative of the types of issues they experienced at the commencement of the project. The data are presented under three broad themes: resourcing, personal challenges, and student challenges. The numbers in brackets indicated which teacher made this comment in their interview. The first two digits represents the school and the second two digits represent the teacher at that school (e.g., 501 is school 5 teacher 1 and 1001 is school 10 teacher 1).

Challenges inexperienced teachers face in very remote contexts (n = 4).

> Teaching in these schools is very very challenging. It is so isolated. There are hardly any resources here, and there is no time at all to make them. We are flat out just coping (501). As a first year teacher it is difficult to know whether I am teaching the concept in the correct way (501). I have no idea how to link the curriculum and my teaching (402), and I don't have the knowledge or confidence that I feel I need to teach mathematics (502). Students' vocabulary is limited. Even adding a new term like 'more' to their language is a hard one for them to grasp (401). I am having to learn their terms just to allow me to understand what they are saying. If they don't have Australian Standard English then they can't be part of the lesson because that is the language I speak to them when I teach (501).

Challenges inexperienced teachers face in metropolitan contexts (n = 4)

Getting resources together is hard. I have to take them home to do (902). It is all very time consuming (702). I am not too sure what to do but as the years go on I am sure I will get better and better (902). I wouldn't say I am fantastic but I try really hard to make sure before the lesson that I really know what I am talking about (905). A lot of these students can't even recognize their numbers (702), and many can only put one or two words together (902). Also their attendance seems to fluctuate (701).

Challenges experienced teachers face in remote contexts (n = 3)

These schools tend to be under-resourced and getting resources to us is very difficult. I am flat out making the resources myself (101). I feel quite confident but sometimes I have to do a little more research and actually 'read' stuff or ask other teachers about stuff (101). I think I am good at Maths language but am not too sure how to run a maths program (102). English is their second language and having that language to them is very important. I model it through focused teaching (201). Attendance is really frustrating for me and the kids. I start something then the kids have got it and I want to move on but can't as kids that haven't attended regularly—I need to reteach them (201). Also there are many kids with behavior problems (202).

Challenges experienced teachers face in metropolitan contexts (n = 5)

The availability of resources is a very big issue for us (901). I am very confident in maths lessons and using the language (802), especially at the year level I am presently teaching (1004). Personally I am not a confident mathematical person (903), but if it is a concept that I have no knowledge about I will find information about it myself before I go and show (301). A lot of the students we have are English as Second Language learners (802, 904). They have limited vocabulary (901). But whatever level they are at in their English I am going to teach what I want to teach and show them I am not going to let their proficiency in Standard Australian English stop them from participating and learning those concepts I want them to learn (302). You have to find different ways of explaining things (301). Most activities must be hands-on and we need to keep recycling the language. There is a lot more repetition and a lot more non-verbal actions (1001).

The difference between beginning teachers and experienced teachers across both contexts as they commenced their RoleM journey was that experienced teachers (a) were more confident about their own understanding of mathematics and teaching mathematics in these contexts, (b) knew where to go if they wanted to know more about a particular mathematical concept, (c) did not see the lack of Standard Australian English as an impediment to learning mathematics, and (d) had in place a range of strategies to help students engage with mathematical concepts. All inexperienced teachers were struggling with all of these dimensions. These struggles were further heightened for inexperienced teachers in the very remote contexts. The difference between the contexts lay in the isolation of the remote and very remote contexts, and the difficulties these teachers had with effectively resourcing their classrooms.

Enacting Mathematics Teaching in These Contexts

Four broad themes emerged from the analysis of the teacher interview data: contextual challenges, teaching mathematics, teacher confidence, and student learning. Each of these themes comprised a range of subthemes. Table 3.4 presents the themes and subthemes that emerged from the data analysis.

Contextual Challenges

The challenges teachers faced working in these communities related to themselves as teachers and their students, with the latter proving to be a greater concern. Student challenges fell into three broad themes: cognitive, affective and contextual, with each consisting of two predominant subthemes. Table 3.5 presents each theme and subtheme with representative quotes from the interview transcripts.

Table 3.6 summarizes the themes and subthemes together with the frequency and percentage frequency of teachers who shared these challenges in each of the three interviews. The final column in the table presents the frequency and percentage

Table 3.4 Themes and subthemes that emerged from the analysis of the interviews

Themes	Sub-themes	
Contextual challenges	Students	Cognitive
		Affective
		Contextual
	Teachers	Pedagogical
		Physical
Teaching mathematics	Pedagogical knowledge	Differentiating the learning
		Teaching maths language
		Engaging students in the learning
		Extending the learning
		Planning and organizing teaching and learning
	Mathematics content knowledge	Mathematical content
		Mathematics language
Teacher confidence		Gains in confidence
		Low teacher confidence
		Confirmed existing confidence
Student learning	Cognitive	Making connections
	Affective	Interest and engagement
		Confidence
		Resilience

Table 3.5 Student challenges these teachers faced in these marginalized contexts

Theme	Subtheme	Representative quote
Cognitive	Language	*Half my class really struggle with the language of mathematics (203, 2011)*
	Prior knowledge	*These students do not even have number recognition (702, 2010)*
Affective	Interest and engagement	*Most of these kids have fairly short attention spans (1602, 2011)*
	Working independently	*Getting kids to work independently [is hard] because I have a lot of lowies (1401, 2012)*
Contextual	Attendance	[*Not*] *coming to school, sometimes we are playing continual catch up (301, 2012)*
	Out of school factors	*If they have problems at home they are going to have problems concentrating (807, 2012)*

Table 3.6 Student challenges identified by the participating teachers (n = 73)

Challenges	Categories	Interview 1	Interview 2	Interview 3	Total/73
Cognitive	Language	35 (47.9 %)	22 (30.1 %)	27 (36.9 %)	59 (81 %)
	Prior knowledge	27 (36.9 %)	16 (21.9 %)	12 (16.4 %)	
Contextual	Attendance	11 (15.1 %)	10 (13.7 %)	7 (9.6 %)	37 (51 %)
	Out of school factors	8 (10.9 %)	3 (4.1 %)	12 (16.4 %)	
Affective	Interest and engagement	19 (26.2 %)	10 (13.7 %)	12 (16.4 %)	34 (47 %)
	Working independently	5 (6.8 %)	7 (9.6 %)	2 (2.7 %)	

frequency of teachers who shared particular challenges (e.g., cognitive, contextual or affective) across the three interviews.

Over the three interviews, 81 % of teachers shared that the cognitive difficulties their students exhibited were associated with their lack of Standard Australian English and prior knowledge. The contextual challenges, that 51 % of these teachers faced, pertained to students' irregular attendance and out of school factors such as home environment. For 47 % of the teachers, the students' lack of interest and ability to work independently was claimed as being of great concern to them. These themes were consistent across the three interviews, the three different geographical locations (metropolitan, remote and very remote), and the three levels of experience (inexperienced, experienced and very experienced). Only the challenges relating to students' prior knowledge and their interest and engagement in learning mathematics decreased as each year progressed.

The personal challenges these teachers faced related to pedagogical impediments (69 % of teachers) and physical impediments (41 % of teachers). The former pertained to catering for students with different ability levels and their own inability to explain ideas to their students. The latter encompassed the lack of physical resources available for teaching mathematics in these contexts and lack of time they had to teach mathematics.

Teaching Mathematics

As acknowledged in the literature, knowledge for teaching mathematics consists of two dominant constructs: pedagogical content knowledge (PCK) and mathematics content knowledge (MCK). The most prevalent construct that teachers shared in their interviews was related to their PCK. Their interviews identified five main sub-themes that changed as a result of their participation in RoleM. These five sub-themes were: knowledge of how to plan and organize their mathematics teaching; knowledge of how to differentiate their teaching; knowledge of how to teach language; knowledge of how to engage their students in learning; and knowledge of how to extend their teaching. Table 3.7 presents the subthemes, dimensions and frequency (percentage frequency) of teachers who shared these dimensions in their interviews. The final column presents that total number of teachers who identified each sub-theme across the three interviews. For example, 60 teachers identified the dimension, *increased knowledge in planning and organizing,* at least once across the three interviews.

The three sub-themes that teachers exhibited the greatest gains in were: planning and organizing mathematics teaching (82 %); new strategies relating to differentiating the learning for their students (82 %); and an improved understanding of how to teach the language of mathematics to their students (70 %). The changes consistently occurred across the year, suggesting that these changes are not instantaneous but require time and ongoing support.

Table 3.7 Dimensions of teachers' 'pedagogical content knowledge' that changed

PCK sub-themes	Dimensions	Interview 1	Interview 2	Interview 3	Total/73
Planning and organizing	Increased knowledge in planning and organizing	34 (46.6 %)	38 (52.1 %)	42 (57.5 %)	60 (82 %)
	Increased use of hands-on resources	19 (26 %)	14 (19.2 %)	13 (17.8 %)	
Differentiating	Learnt new teaching strategies	44 (60.3 %)	36 (49.3 %)	37 (50.8 %)	60 (82 %)
Teaching language	Improved understanding of teaching language	23 (31.5 %)	24 (32.9 %)	22 (30.1 %)	51(70 %)
	Increase in students' exposure to language	0 (0 %)	5 (6.8 %)	7 (9.6 %)	
Engaging	Increase in students' engagement	14 (19.2 %)	11 (15.1 %)	9 (12.3 %)	26 (36 %)
Extending	Improved understanding of extension	11 (15.1 %)	10 (13.7 %)	6 (8.2 %)	22 (30 %)

Table 3.8 Sub-themes of teachers' 'mathematics content knowledge' that changed

MCK	Sub-themes	Interview 1	Interview 2	Interview 3	Total/73
Math content	Teacher MCK increased	13 (17.7 %)	21 (28.7 %)	26 (35.6 %)	52 (71 %)
	Teacher MCK activities increased	5 (6.8 %)	4 (5.5 %)	3 (4.1 %)	
	Teacher MCK clarified and changed	5 (6.8 %)	4 (5.5 %)	8 (10.9 %)	
Language	Learning of mathematical language	9 (12.3 %)	9 (12.3 %)	12 (16.4 %)	27 (37 %)

Table 3.9 Gains in teacher confidence

	Interview 1	Interview 2	Interview 3	Total/73
Gains in teacher confidence	37 (51 %)	41 (56 %)	58 (80 %)	69 (95 %)
Low teacher confidence	28 (38 %)	10 (14 %)	9 (12 %)	35 (48 %)
Confirmed existing confidence	7 (10 %)	3 (4.1 %)	9 (12 %)	17 (23 %)

The gains that teachers made with regard to their MCK fell into three broad sub-themes: increased MCK in general; increased MCK with regard to using MCK knowledge to breakdown and structure effective activities for the classroom; and MCK was clarified and changed. Table 3.8 presents the sub-themes of teachers' MCK that changed over the three interviews.

The gains made in MCK significantly increased from interview to interview with the greatest emphasis being placed on the teachers' own mathematics content knowledge.

One issue that continually emerged in the data related to teachers' confidence in teaching mathematics. Ninety-five percent shared that their confidence in teaching mathematics in these contexts and to these marginalized students had increased due to their participation in RoleM professional learning. Table 3.9 presents the gains teachers made with regard to their confidence.

Even though nearly all teachers shared that their self-confidence had improved, almost half the sample still disclosed that they were under-confident with regards to teaching mathematics.

Student Learning Gains

The last theme that teachers shared in their interviews related to the gains that they saw in their students' learning across the course of the RPL. This theme consisted of two predominant sub-themes: cognitive gains and affective gains. Table 3.10 presents the sub-themes, the dimensions of each subtheme and representative quotes from the interview data. Table 3.11 presents the frequency and percentage frequency of teachers who referred to these dimensions across the three interviews.

Table 3.10 Representative teacher quotes for the subthemes of student learning

Sub theme	Dimension	Representative quote
Cognitive	Making connections	*They were playing in the new sand pit the other day and were using language like who has the biggest/smallest etc. (301, 2010)*
Affective	Engagement/Enjoyment	*They are happy to write things down on paper because they have done it first. They understand it (1310 2013)*
	Confidence	*They are more confident. I definitely know they can do it, so I have higher expectations of them (1210, 2011)*
	Resilience	*I am seeing them develop into independent workers which I didn't even think was possible (303, 2011)*

Table 3.11 Dimension of student learning that teachers believed changed

	Interview 1	Interview 2	Interview 3	Total/73
Making connections	15 (21 %)	37 (51 %)	33 (45 %)	55 (75 %)
Engagement/Enjoyment	39 (53 %)	46 (63 %)	37 (51 %)	61 (84 %)
Confidence	1 (2 %)	15 (20.5 %)	13 (18 %)	32 (44 %)
Resilience	3 (4 %)	7 (9.6 %)	2 (3 %)	10 (14 %)

Thus, the greatest change that teachers saw in their students was their increased engagement in the mathematics classroom. They persistently shared these changes across all three interviews. They also evidenced that students were making connections between different mathematical concepts, both in school and out of school contexts. This became more evident as the year progressed.

In order to gain some understanding of the changes that had occurred due to the RoleM professional learning, the final interviews with the 16 teachers from 2010 (the teachers who shared beginning at RPL stories) were analyzed to ascertain just what it was like for them at the conclusion of the RPL. The next section shares these 'end journey' stories.

At the Conclusion of One Year of RoleM Professional Learning

Changes inexperienced teachers made in very remote contexts (n = 4).

> After teaching out here in these schools a while it is getting better. I think the [RoleM] PL just reminded me of the importance of getting in there with the students and play[ing] with them (401). The kids are engaged and they don't even realize they are learning, if you have something that is play based with a lot of concrete materials they get really engaged (502). I suppose my teaching has changed in that I am a first year teacher so I don't have

experience with oral language specific[ally] in mathematics. I am [now] more educated and I know what language to use in my classroom. I really gained a lot of knowledge on how language is the focus for all teaching mathematic concepts (501). [I am now more capable of] using the activities to build on students' mathematical language and using that actually in play situations (501). It [RoleM] gives me more confidence knowing what activities can be provided for the students [and] linking them to the curriculum (402).

Changes inexperienced teachers made in metropolitan contexts (n = 4)

It has changed and improved my understanding of mathematical concepts, basically the skeleton of the concepts and be[ing] able to break it down, so I can pass it on (701). I just like getting the things [RoleM resources] and learning how to use some of the things until you feel confident (902). Different ideas of how to teach concepts, so many ideas on how to do things and help make mathematics become fun for kids (905). From day one they have been engaged (702). Kids that started the year with me are doing really well, compared to last year, I understand more of what I want them to do; it is not just counting, it is number knowledge. I do question [them more] and I do push them a lot more [now] (701).

Changes experienced teachers made in remote contexts (n = 3)

The kids are improving and am seeing evidence, you just have to listen to them talk and the things they are giving back to you. I recognize there is a language difference, but I don't dumb it down [anymore]. They need to know these words (101). It made me look at certain things I was doing and made me a bit critical of what I was doing before. A lot of times I haven't formalized things as much as I should have. I am doing formal lessons, I am being more specific in what I am looking at and getting a chance to talk to the kids about how they think about those things (102). When you do focused teaching and learning, having those [RoleM] activities allows students to play with them. Students get excited about the maths. Their understanding is better. You can integrate it into lots of incidental learning. You can put it on the table and only a few might get interested, then you bring the whole class in...If it's coming from them, they seem to retain it more, they want to engage in it more (201). It has given me more confidence to teach math. The little activities it gives more confidence to teach, when they have little games you notice that the activities they are doing in maths come out (202).

Changes experienced teachers made in metropolitan contexts (n = 6)

I have got kids that are at all of those levels in my class. So it was really good for me to see how I can integrate all that across learning: to make sure all kids are engaged and meeting outcomes. They [students] are a lot more confident in their ability and are prepared to have a go. Some were hesitant in the beginning (301). It [RoleM] challenged me to do more maths and engage my students and not be airy-fairy and be direct in my teaching. My expectations [for my students] have risen (302, 802). Even if they are away, they will come back and have a go. Maths is enjoyable in my class now (302). I probably use a lot more mathematical language. It's a lot more explicit, there are a lot more different terms. Instead of using the same language all the time, like more or less, using more terms so the kids are getting more variety as well (802). I am using the language more consciously (901), and am more confident to use the actual terms (1004). They're starting to mimic my language so you can see their—like in the beginning of the program you'd ask them a question and you'd get a word answer but now it'd be more—'well three comes after two but it's also before four' (802). They are a bit more confident in their own abilities to use mathematical language correctly and understand the concepts (904). Even positional language they're doing a lot better (1001). Whereas before it was 'I don't know if I can do that'. Now I say 'right show me this'. I am more confident in what I am asking the children (1004).

Trends Across the Themes

An analysis of the themes and sub-themes delineated in the above sections gave insights into the order of importance teachers gave to each. The more frequently they mentioned a sub-theme the more important that sub-theme was considered to be. For example, the sub-theme that most teachers shared was *gains in their teacher confidence* (95 %). For the purpose of this analysis it was decided that if more than 70 % of teachers referred to a particular theme and subtheme across the three interviews then this theme and sub-theme was significant. In all there were eight themes that were considered significant for these teachers. Table 3.12 presents subthemes and trends in order of importance teachers placed on each.

The area where teachers and students made greatest gains was in the affective domain, with teachers' confidence in teaching mathematics and students' engagement and enjoyment in learning mathematics significantly increasing as a result of their participation in RoleM. The latter was closely followed by an increase in teachers' pedagogical knowledge, particularly their understanding of how to differentiate the learning for students in their classes, and how to plan and effectively organize their teaching of mathematics. From these teachers perspective, the challenges that that they consistently experienced with their students related to their inability to communicate in Standard Australian English and their lack of prior knowledge of mathematical concepts. Finally, teachers shared that their own mathematics content knowledge and understanding of how to teach mathematical language and ways of communicating mathematically to these students had significantly increased over the year that they participated in the RoleM project.

In order to gain insights into the consistency of these themes and sub-themes across different contexts, the data were split according to teachers' geographical location and teachers' experience. Tables 3.13 and 3.14 present the results of these analyses. The shaded cells indicate the cohorts of teachers where the percentage of teachers fell below 70 %.

Table 3.12 Emergent subthemes in order of importance

Theme	Sub theme	Trend	Agreement (%)
Teacher self confidence		Increased	95
Student learning	Engagement/Enjoyment	Increased	84
PCK	Planning and Organizing	Increased knowledge	82
PCK	Differentiating	Learnt new strategies	82
Student cognitive challenge	Language and prior knowledge		81
Student learning	Making connections	Increased	75
MCK	Teachers MCK	Increased	71
PCK	Teaching language	Improved understanding	70

Table 3.13 Agreement to the themes by context

Theme	Metropolitan (n = 42)	Remote (n = 19)	Very remote (n = 12)
Teacher self confidence (gains)	40 (95 %)	18 (95 %)	11 (92 %)
Student learning (engagement)	34 (81 %)	16 (84 %)	11 (92 %)
PCK Differentiating	35 (83 %)	12 (74 %)	11 (92 %)
PCK Planning and Organizing	35 (83 %)	15 (79 %)	10 (83 %)
Student challenges (cognitive)	34 (81 %)	16 (84 %)	9 (75 %)
Student learning (making connections)	**29 (69 %)**	15 (79 %)	11 (92 %)
Teachers MCK (gains)	32 (76 %)	**11 (58 %)**	9 (75 %)
PCK Teaching language (gains)	**29 (69 %)**	12 (74 %)	**7 (58 %)**

Table 3.14 Agreement to the themes by experience

Theme	Inexperienced (n = 38)	Experienced (n = 19)	Very experienced (n = 16)
Teacher self-confidence (gains)	36 (95 %)	18 (95 %)	15 (94 %)
Student learning (engagement)	32 (84 %)	16 (84 %)	13 (81 %)
PCK Differentiating	34 (90 %)	15 (79 %)	**11 (69 %)**
PCK Planning and Organizing	31 (82 %)	15 (79 %)	14 (88 %)
Student challenges (cognitive)	30 (79 %)	16 (84 %)	13 (81 %)
Student learning (making connections)	29 (76 %)	15 (79 %)	**11 (68 %)**
Teachers' MCK (gains)	30 (79 %)	**11 (58 %)**	**11 (68 %)**
PCK Teaching language (gains)	28 (74 %)	15 (79 %)	**8 (50 %)**

In addition to these themes, teachers from remote locations (n = 19) also shared that the contextual challenges relating to their students (e.g., absenteeism, etc.) were of significance (74 %).

It seems that the gains in confidence that teachers made with regard to teaching mathematics was consistent across all three levels of experience. Even the most experienced teachers shared that their confidence had strengthened as a result of their participation in RoleM. All three groups also reported that their PCK with regard to planning and organizing the delivery of mathematics activities had also increased.

In summary, the three predominant themes and subthemes that teachers shared in their interview data across their year of participation in RoleM in descending order of agreement were:

- gains in their self confidence (92 %)
- gains in their PCK—planning and organizing and differentiating (new teaching strategies (82 %))
- gains in their MCK (75 %)

The order of these themes was consistent across all geographic locations and years of teacher experience. There were also shifts in the degree of agreement to

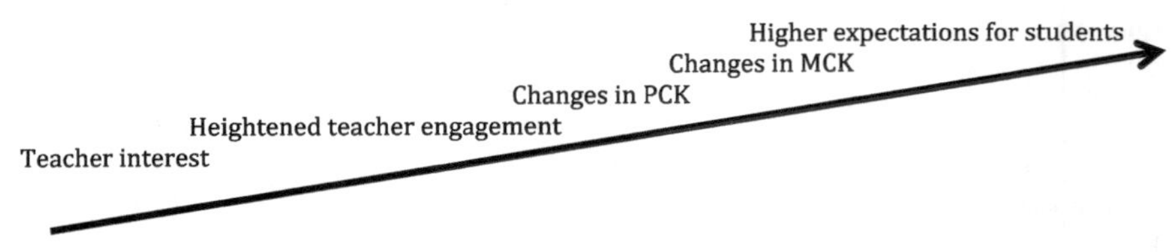

Fig. 3.1 Emerging professional learning trajectory

these subthemes as the RoleM project moved across the first three years of the longitudinal study.

To gain insights into these shifts, an in-depth analysis of the interviews, 12 teachers from one school who had participated in RoleM over the first three years of the longitudinal study were examined in depth (see Warren and Miller 2013 for complete analysis of these shifts). In summary these shifts were:

PCK: In the first year of RoleM, teachers had little to say about how their teaching practice (PCK) changed across the whole year. Their focus was predominantly on the personal gains they had made with regard to their increased confidence and increased enjoyment in teaching mathematics. By contrast, teachers from the second and third year of the project were much more explicit with regard to how their practice had changed, and towards the end of each year began to discuss how they were now setting higher expectations for their students.

MCK: With regard to MCK, the first year cohort did not appear to engage with the mathematical content until their second interview, and then they progressed from mathematical language (Interview 2) to mathematical content knowledge (Interview 3). By contrast, the second and third year cohort began the year with a focus on mathematical language and mathematical content knowledge, and sustained this throughout the year. Emerging from this analysis, a professional learning trajectory with the stages that teachers' progress through as they become experienced in teaching mathematics in disadvantages contexts was delineated (see Fig. 3.1). This was:

This trajectory aligns with the stages identified in our previous research with regard to supporting teachers to become effective teachers in disadvantaged contexts (Warren and Quine 2013), namely, building teachers' confidence, building students' confidence (resulting in changes in teachers' PCK), and finally increasing expectations for students (resulting in changes in teachers' MCK). This trajectory is discussed further in Chap. 6.

Dimensions of the RoleM Professional Learning that Supported These Teacher Changes

Throughout the interviews teachers also shared the strategies embedded in the RPL model that assisted them on their journey of change. The two dimensions of the

Table 3.15 Most helpful strategies in supporting teachers' changed practices

Dimension	Strategy	Interview 1	Interview 2	Interview 3
Dialoguing with a knowing person (PD)	Provision of ready-made learning activities	70 (96 %)	67 (92 %)	65 (85 %)
	Provision of ready-made resources	47 (64 %)	25 (34 %)	27 (37 %)
Collaborative implementing and sharing (in-situ)	Demonstrating the implementation of the learning activities	18 (25 %)	38 (52 %)	18 (25 %)
	Demonstrating how to use the ready-made resources	8 (11 %)	9 (12 %)	18 (25%)

RoleM professional learning model that the majority of teachers were extremely positive about were *dialoguing with a knowing person* and *collaborative implementing and sharing*. Table 3.15 presents the strategies that they identified as being most helpful in supporting their changed practices, together with the frequency and percentage frequency of teachers who identified these in their interviews.

The theme that consistently occurred across all three interviews was the provision of ready-made learning activities. It should be noted that these learning activities were simply not given to the teachers to implement. The focus of the PL days that occurred prior to the implementation of these learning activities in the classroom entailed detailed discussions and modeling of how these learning activities could be enacted in their teaching. In addition, when the team had in-situ visits within the teachers' classrooms teachers also had the opportunity to request that the RoleM team demonstrate how they would implement learning activities. On the whole they found the PL days much more helpful than the in-situ visits.

Dialoguing with a Knowing Person

The two strategies that made a substantive impact on these teachers' teaching of mathematics were the provision of the ready-made learning activities and the ready-made resources. To understand more fully just how they assisted these teachers, representative quotes relating to each of the dimensions are presented in the next section. Each representative quote is concluded with two numbers in brackets (e.g., 1501, 2011). The first number is the number allocated to each teacher who was interviewed. The second number indicates the year in which this interview occurred. Thus (1501, 2011) indicates that teacher 01 at school 15 made this comment in 2011.

The first action that positively impacted on teachers was the enacting of the RoleM learning activities (learning activities that they would be implementing in their classrooms) using hands-on materials.

> I [now] realize how much I have appreciated this PD [Professional development] and how diverse it has been. I love being able to get out and just sit down and just do really hand-on activities. It helps you to remember it more as well 'cause I guess that teachers also like to learn by doing (1501, 2011). It was practical and got me thinking how I could use it. More practical than theory (302, 2010).

It seems that this action reminded them that hands-on doing helped them to learn, and in turn this also helps their students to learn.

The second action was the movement through the different levels of learning embedded in each learning activity. Each RoleM learning activity consisted of three different levels of learning (consolidating, engaging and extending). As teachers 'acted out' each activity the learning that occurred at each level was discussed together with the accompanying mathematical language each required.

> I liked breaking down the concepts and the different activities… the activities involve logical thinking, especially when we break it down and [ask] logical questions. It helped [me learn] about mathematical language (203, 2011). It gave more ideas [of how] to really focus on the ways students learn. I think it just reminded me of the importance of getting in there with the students (401, 2010). … and help students gain a better understanding (501, 2010). It brought me back to ok what are the strategies that children need to learn to be able to do these problems. Gave me ideas to do extension and to be able to provide different levels of work (108, 2013).

Not only did this second action result in an increase in their mathematical knowledge but it also caused them to return to their core role, helping students to understand and using strategies that can best assist this occurring.

The third action was the inclusion of directed activities that focused on the core concepts of mathematics with resources that visually made each concept very apparent to the learner.

> …sort of see how you're teaching and [how] to take that to a different place (1703, 2013). It challenged me because I know I wasn't doing enough mathematics… challenged me to engage my students and not do airy fairy stuff in maths. It has given me more direction (302, 2010).

Collaborative Implementing and Sharing (In-Situ)

The two actions that most assisted teachers in the implementation component of the RoleM professional learning model were: (a) modeling activities in their classroom that they found difficult to implement; and (b) allowing teachers time to 'watch' and 'reflect' on what their students were doing as the lessons were being modeled.

> Demonstration of how to use them [the resources] in the class, it is one thing to know how to do it, but to see it in action is much better (104, 2011). Reading the book I am so overwhelmed and I am like 'what is this all about' (207, 2012). I didn't know what I am was supposed to do (911. 2012) and then once it was modeled it clicked in my head and I thought oh that makes sense (907, 2011).

While these teachers had themselves modeled the implementation of the learning activities in the PD, it seems that the transfer of these actions to a classroom situation is not necessarily an easy process. As one teacher shared, *There's a lot to get through in one day... a lot goes in one ear and out the other (911, 2012).* It seemed that seeing the learning activity 'in action in situ' was a key practice that helped them engage in the implementation stage.

> It also gave a chance to walk around and observe kids that I might not get to see as much during a maths lesson… to see how lessons are done in real life application [helped] me change what I do (305, 2013). Just because we are so remote it is good to see ideas and how they would be implement and … just to have people come out and throw ideas around and just how to 'run' with the kids… seeing other kids interacting with other people and being able to apply the knowledge I had taught them with other people (154, 2013).

The action of allowing teachers the time to walk around the classroom and observe the implementation of the activities with their students helped them to (a) change the pedagogy they used in their classroom, and (b) observe what their students knew. In some instances this was affirming for them as teachers and in some cases it was surprising.

In three instances, the model used for collaborative implementing and sharing was modified. These changes included: the number of members of the RoleM team who went to work in the classrooms was increased to five, and the activities they used were different from those recommended in the RoleM learning activities or at the professional development. The participating teachers were very vocal with regard to their dislike of these changes. The reasons they gave were:

> I was really disappointed. I really wanted to see her model on from the book to make sure my interpretation of the activity is the same as hers (907, 2011)… because you have four teachers in there it is not a realistic model (1010, 2012). We live so far away. It does give us good ideas but I don't know where to go for the resources (1703, 2012).

This feedback was a reality check for the RoleM team. While these changes were perceived by the RoleM team as making the experience 'more exciting' for students (and the participating teachers), it is important to ensure that classroom modeling utilizes resources that teachers have ready access to.

Learning Activities and Resources

Throughout the interviews teachers also shared the aspects of the learning activities and resources that best supported them in their teaching of mathematics in marginalized contexts. These fell into two broad areas: the physical attributes of the resources themselves;

> They [the resources] are ready to go so it was up to me to organize when I wanted to do everything… use it with a small group or a whole class on the mat (302, 2010). They [the materials] are engaging and concrete (501, 2010)…

and, the pedagogical and content knowledge embedded within and across the learning activities.

> [Showed me] the importance of teaching concepts in lots of different ways. For example using lots of different resources and representations to each numbers to 100. Not just the hundreds board. That way they can transfer their knowledge. (804, 2011).... What questions to ask, strategies to use, and how to modify the games [activities] to extend the children. Getting away from structured teaching (501, 2010).

The data presented in this chapter offer some insights into how a well-designed professional learning model can assist and support the transformation of teachers' pedagogical practices, their knowledge of mathematics, and their beliefs about themselves as teachers of mathematics. As evidenced by the data, this transformation is complex, requiring more than 'one off' discrete training sessions. The development of appropriate knowledge of mathematics and pedagogical practices required to teach students in these marginalized contexts is more likely to occur when an iterative cycle of reflection, collaboration, ongoing support and feedback, and professional learning is more intensive and of longer duration. The final section in this chapter shares the results relating to the Indigenous teacher aides, who participated in RoleM.

The Inclusion of Indigenous Teacher Aides in RoleM

Background

Internationally, there is a concern for improving educational outcomes for Indigenous students (DEST 2007), and the underperformance of Indigenous Australian students is well documented (OECD 2014). In schools with large Indigenous student enrolments a common strategy used to redress this issue is the employment of Indigenous Teachers and Indigenous Teacher Assistants (ITA) (DEST 2007). While in the Australian school context the employment of ITAs has been occurring from the early seventies, their voice in educational initiatives and educational research aimed to best support Indigenous student learning is almost nonexistent. Given that the discourse used in the educational arena can in fact further marginalize Indigenous students, particularly discourse of comparison with non-Indigenous students (Harrison 2007), the lack of representation of ITAs in the literature is concerning.

While ITAs experience little power in the education arena, they are in a unique position in that they can be the link between the two cultures and two knowledges. ITAs 'walk' and work between the two knowledge systems of Indigenous knowledge and Western knowledge in Australian schools (Fitzgerald 2006). ITAs are aware of Indigenous knowledge and cultural issues that non-Indigenous staff are not privileged to know. They also have to address the Western knowledge that is incorporated by the school (MacGill 2008). Thus ITAs act as a buffer for both Indigenous and

Western knowledge systems in the school community (MacGill 2008). To begin to address this problem, Indigenous educational leaders suggest schools adopt a two-way learning approach for Indigenous students (Pearson 2009; Sarra 2011; Yunupingu 1991). These ITAs are usually employed on a full time basis and one is assigned to each classroom. Their role is to 'work' between the two knowledge systems, Indigenous and Western, to cross between the two cultures. They are aware of knowledge and culture issues that non-Indigenous teachers have no access to.

Research indicates that non–Indigenous teachers in Indigenous settings dominate the learning environment. Warren et al. (2004, 2010) found in their study of the interactions between non-Indigenous teachers and Indigenous teacher assistants in remote Australian schools, that of the teachers interviewed, not one acknowledged an equitable partnership with their ITA. Teachers believed their role was to direct the ITA and determine the duties they required to be conducted in their classroom (Warren et al. 2009). These duties commonly consisted of providing administration assistance, managing classroom behavior, and, tutoring individual students who were identified as risk learners (Warren et al. 2009). Evidently, teachers control their own classroom environments and provide little opportunity for ITAs to have an input and ownership of students' learning. Warren and colleagues (2004, 2010) reported that very few ITAs were involved in planning or asked to contribute their ideas on student learning, for example Indigenous learning styles, cultural awareness and students' backgrounds.

While there has been an ongoing emphasis on providing opportunities for professional learning and upgrading qualifications for ITAs, the options are varied and in some cases seen as not fulfilling ITA's educational needs. In addition, many ITAs feel intimidated in taking up educational opportunities (Yates cited in Winkler 2006). In Australia, most of the ITA programs currently available are located in remote areas (York and Henderson 2003), they appear to not provide professional learning in areas such as, conflict resolution, effective communication, information technology, and up-skilling ITAs in literacy and numeracy pedagogy (Yates cited in Winkler 2006). Thus, it is perceived that the type of training presently provided to ITAs is not equipping them with the skills required to be successful in their role in the classroom assisting students to learn (Yates cited in Winkler 2006). Due to the limited number of programs offered to assist ITAs in furthering their educational pursuits, most ITAs access professional development sessions through schools and educational consultants. These sessions often focus on key learning areas from the curriculum and are broad in their delivery. While this is beneficial the gap still remains where ITAs are receiving little training in how to assist teachers in educating Indigenous students (Cooper et al. 2005; Winkler 2006).

Strategies to Help Support These ITAs

Drawing on past research findings with regard to the role and position of ITAs in classroom contexts, specific strategies were utilized in RoleM with the aim of strengthening the partnership between the ITAs and their teachers. Briefly these were:

- Funding the ITAs to attend the professional learning workshops alongside their classroom teacher. When teachers and ITAs work in collaboration Indigenous student learning improves (Malloch 2003).
- Using the ITAs as facilitators during the implementation sessions in the classroom. A way of ensuring that the ITAs role in the classroom incorporates 'active' participation in the learning of their Indigenous students was to assign them to a specific role in the implementation of the RoleM learning activities. Two-way learning requires teachers and ITAs to be 'equal' collaborators' (MacGill 2008) and to have ITAs more involved in the educational outcomes of Indigenous students (Warren et al. 2004).
- Supporting ITAs who wanted to participate in further study. This included helping them to identify their goals and aspirations, assisting them to enroll in their chosen course, and tutoring and mentoring them as required.

Participants

Ten of the 16 schools that participated in RoleM had a large proportion of Australian Indigenous students, with three of these 10 schools being fully Indigenous. Eight were located in remote and very remote regions and two in metropolitan regions. Nineteen ITAs participated in three interviews in their year of participation with RoleM. The interviews were designed to investigate their beliefs and attitudes, and understandings with regard to (a) their role in the classroom, (b) their perspectives on Indigenous student learning, and (c) their perspectives of what prevents Indigenous students from effectively participating in mathematics. The next section shares the results of these interviews with accompanying representative quotes that provide insights into the nature of the three themes that emerged from the analysis of the interviews: changed role; Indigenous student learning; and factors that Indigenous students learn.

Results

Changed Roles

Nearly all ITAs felt that as a result of their participation in RoleM they were more confident in teaching mathematic and more involved in facilitating mathematics learning.

> It [RoleM professional learning] makes me more confident to do what I am supposed to do and take a group singularly (ITA04). Feels like I have a more active role in helping students learning (ITA11).

This movement to working with small groups, not only helped them plan their teaching experiences for groups of students, but also gave them greater opportunities to relate students learning to their cultural background. For example, as one ITA noted, *if they don't understand it we just use something else as an example or change pictures/resources to familiarize* [*students to*] *it* [*the concept*].

> Being Indigenous, how would I say that to my own kids so they get it easier. I may have to break it down instead of two or three instructions at once. I will say you need to do this and then come back. Now you need to..... and so on (ITA03).

Indigenous Students Learning

The three main aspects of learning that they shared with regard to their Indigenous student learning during the implementation of RoleM related to their affective domain.

> (a) They are engaged and once they get it they are so excited. (ITA03); (b) they seem more confident and they are willing to explore. When we expose them to the RoleM stuff they seem to get more into it (ITA12); and, (c) Before RoleM I had to break things down easily as I could so the kids got frustrated. But now the kids don't see maths as bad, it is a log more fun now and its not putting stress on the kids (ITA09)

Teachers' perspectives of ITAs

Teachers indicated their working relationships with ITAs changed after attending the RoleM PL. Teachers shared that the professional learning sessions with ITAs helped ITAs to become more involved in classroom planning, routine and curriculum in mathematics. In addition, ITAs were now able to continue with the supporting student learning when teachers were absent. A direct consequence of this change in role was limiting disruptions in learning for students. A teacher from North Queensland reflected:

> This term I gave her (ITA) activities to do and the next day I could not come and it was really great that she did the activities with the students. She knew what she was doing. We were in groups today and she knew what to ask, like "how many repeats can you see?" She was implementing those questions. (T205, I2, 2011, L 69–74).

The way teachers utilized ITAs in the classroom also changed. The focus was taken off behavior management and concentrated on using them in more effective roles that helped students learn.

> We are using our teacher aide (ITA), Tina more effectively as well, not just having her in the background controlling behavior while I am trying to teach. Now we are doing rotations every week, she is now allocated a group so she has the responsibility of working with the children rather than just supporting me. She (ITA) knows how to do everything; she is in there doing it. (T102, I2, 2011, L50–57).

Discussion and Conclusion

The outcome for ITAs participating in RoleM professional learning was a very positive experience for both: (a) establishing their own role and identity in students mathematical learning, and (b) creating equitable partnerships between teachers and ITAs. While not reported in this paper, it is essential to highlight that students (both Indigenous and non-Indigenous) made positive gains in their mathematical learning (see Warren and Miller 2013). Measuring the effectiveness of the program was essential to demonstrating the success of the professional learning, which past research has highlight is challenging to implement across large school reform projects for Indigenous students (Bishop 2012). These positive gains were evident in all geographical locations across all year levels that participated in the study.

The inclusion of ITAs in professional learning is the first step towards developing an equitable relationship between the teacher and the ITAs. The influence that the RoleM strategies had on the teacher—ITAs partnership was profound. The results confirm and add to Malloch's (2003) study, that is, including ITAs in the planning and working with small groups assisted to strengthen the effective partnership between teachers and ITAs. An effective partnership teacher—ITA partnership involves having ITAs actively contributing to the students' learning, with a clearly delineated task for them to do within the groups. It also requires that they are confident in and value the task that has been set for them, and take on the ownership of student learning. This shift in role allowed ITAs to be involved in not just the cultural side of student learning but also the pedagogical side, a shift that rarely occurs in many of these contexts (Warren et al. 2009).

ITA and teacher partnerships were central to continued authentic engagement and classroom learning experiences (Armour et al. in press). ITAs played a major role in moving between the two worlds (Western and Indigenous) to assist students to access and engage with the mathematics. Additionally, they worked in conjunction with the teachers to facilitate best practice to support student learning. The strengthening of this partnership ensured that the implementation of the project was successful at all sites. On large-scale projects consistency of implementation across schools and within schools is challenging (Bishop 2012). We conjecture that as an equitable relationship developed over the course of the RoleM PL ensured continued consistent delivery of the project as both teachers and ITAs were accountable for the student learning. It is also acknowledged that while this approach has very positive outcomes it requires time and resources.

References

Armour, D., Warren, E., & Miller, J. (in press). Working together: Strategies that support two-way learning in the mathematics classroom. *Asia-Pacific Journal of Teacher Education*.

Aubrey, C. (2003). When we are very young: The foundations for mathematics. In I. Thompson (Ed.) *Enhancing primary mathematics teaching* (pp. 43–53).

Ayalon, A. (2011). *Teachers as mentors: Models for promoting achievement with disadvantaged and underrepresented students by creating community*. Stylus Publishing: Sterling, Virginia.

Bishop, R. (2012). Pretty difficult: Implementing kaupapa Maori theory in English-secondary medium schools. *New Zealand Journal of Educational Studies, 47*(2), 38–50.

Cooper, T. J., Baturo, A., & Warren, E. (2005). Indigenous and non-indigenous teaching relationship with three mathematics classrooms. In H. Chick & J. Vincent (Eds.), *Proceedings of the 29th Annual Conference of the International Group for the Psychology of Mathematics Education* (Vol. 2, pp. 265–273). Melbourne, VIC: PME.

Darling-Hammond, L., & Post, L. (2000). Inequality in teaching and schooling: Supporting high-quality teaching and leadership in low-income schools. In Richard D. Kahlenberg (Ed.), *A notion at risk: Preserving public education as an engine for social mobility* (pp. 127–167). New York: The Century Foundation.

Department of Education, Science and Training. (2007). *What works: Indigenous Education*. Retrieved from http://www.whatworks.edu.au/upload/1311202790052_file_CoreIssues7.pdf.

Feiman-Nemser, S. (2001). From preparation to practice: Designing a continuum to strengthen and sustain teaching. *Teachers College Record, 103*(6), 1013–1055.

Fitzgerald, T. (2006). Walking between two worlds: Indigenous women and educational leadership. *Educational Management Administration and Leadership, 34*(2), 201–213.

Frigo, T., Corrigan, M., Adams, I., Hughes, P., Stephens, M., & Woods, D. (2004). Supporting English literacy and numeracy learning for Indigenous students in the early years. *ACER Research Monograph* (Vol. 57). Melbourne: ACER.

Harrison, N. (2007). Where do we look now? *The Future of Research in Indigenous Australian Education, Australian Journal of Indigenous Education, 36*, 1–5.

Hill, H. C., Rowan, B., & Ball, D. L. (2005). Effects of teachers' mathematical knowledge for teaching on student achievement. *American educational research journal, 42*(2), 371–406.

Hoy, A. W. (2000, April). *Changes in teacher efficacy during the early years of teaching*. Paper presented at the annual meeting of the American Educational Research Association, New Orleans, LA. Retrieved from http://www.coe.ohio.state.edu/ahoy/efficacy/html.

Jurdak, M. (2009). *Toward Equity in Quality in Mathematics Education*. Boston, MA: Springer.

Lupton, R. (2004). *Schools in Disadvantaged Areas: Recognising Context and Raising Quality*. Retrieved from http://sticerd.lse.ac.uk/dps/case/cp/CASEpaper76.pdf.

MacGill, B. (2008). *Aboriginal Education Workers in South Australia: Towards equality of recognition of Indigenous ethics of care practices* (Unpublished doctoral dissertation) Flinders University, Bedford Park, South Australia.

Malloch, A. (2003). Teacher and tutor complementation during an in-class tutorial program for Indigenous students in a primary school. In S. McGinty (Ed.), *Sharing success an Indigenous perspective: Papers from the second National Australian Indigenous education Conference* (pp. 181–204). Altona, VIC: Common Ground Publishing.

Organisation for Economic Co-operation and Development (OECD). (2014). *PISA 2012 Results: What Students Know and Can Do*. Vol. I (revised edition).

Organisation for Economic Co-operation and Development (OECD) (2012) *Equity and quality education: Supporting disadvantaged students and schools*. Retrieved from 10.1787/9789264130852-en.

Pearson, N. (2009). Radical hope: Education and equality in Australia. *Quarterly Essay, 35*, 1–105.

Sarra, C. (2011). *Strong and Smart: towards a pedagogy for emancipation: education for first people*. London; New York: Routledge.

Smith, J. A., Flowers, P., & Larkin, M. (2009). *Interpretative Phenomenological Analysis: Theory, Research, Practice*. London: Sage.

Strauss, A., & Corbin, J. (1990). *Basics of qualitative research: Grounded theory procedures and techniques*. Newbury Park, CA: Sage.

Strauss, A., & Corbin, J. (1998). *Basics of qualitative research: Techniques and procedures for developing grounded theory* (2nd ed.). Newbury Park, CA: Sage.

United Nations Educational, Scientific, and Cultural Organisation (UNESCO). (2014). *Teaching and learning: Achieving quality for all.* Paris: Author.

Warren, E., Baturo, A., & Cooper, T. (2010). Power and authority in school and community: Interactions between non-Indigenous teachers and Indigenous teacher assistants in remote Australian schools. In J. Zadja & M. A. Geo-JaJa (Eds.), *The Politics of Education Reforms* (pp. 193–207). Dordrecht, The Netherlands: Springer.

Warren, E., Cooper, T. J., & Baturo, A. (2009). Bridging the educational gap: Indigenous and non-Indigenous beliefs, attitudes and practices in a remote Australian school. In J. Zajda & K. Freeman (Eds.), *Race, Ethnicity and Gender in Education* (pp. 213–226). New York, NY: Springer.

Warren, E., Cooper, T., & Baturo, A. (2004). Indigenous students and Mathematics: Teachers' perceptions of the role of teacher aides. *Australian Journal of Indigenous Education, 33*, 37–46.

Warren, E., & Miller, J. (2013). Enriching the professional learning of early years teachers in disadvantaged contexts: The impact of quality resources and quality professional learning. *Australian Journal of Teacher Education, 38*(7), 91–111.

Warren, E., & Quine, J. (2013). A holistic approach to supporting the learning of young indigenous students: One case study. *Australian Journal of Indigenous Education, 42*(1), 12–23.

Winkler, M. (2006). Hidden treasures: Recognising the value of Indigenous educators. *Education Horizons, 9*(2), 18–19.

York, F. A., & Henderson, L. (2003). Making it possible: The evolution of RATEP—A community based teacher education program for Indigenous people. *The Australia Journal of Indigenous Education, 23*, 77–84.

Yunupingu, B. (1991). A plan for Ganma research. In R. Bunbury, W. Hastings, J. Henry, & R. McTaggart (Eds.), *Aboriginal pedagogy: Aboriginal teachers speak out* (pp. 98–106). Geelong: Deakin University Press.

Chapter 4
Marginality and Mathematics

Abstract As evidenced in Chap. 3, the problem of disadvantaged students underperformance in mathematics can be addressed by changing how mathematics is enacted within the classroom context. This chapter investigates the issue of disadvantage in terms of the context in which students live, their cultural and educational background and the impact these have on learning mathematics (what these students bring to the equation). Its particular focus is on the participating students and their development as they move through the four years of the longitudinal study. At the completion of the project, the majority of students' achievements in mathematics mirrored that of their mainstream counterparts. Students' growth is measured according to their pre and post-test scores on a test administered at the commencement and conclusion of each year. This growth is examined in terms of students' geographical location, ethnicity, and teacher experience. In addition, the chapter includes illustrative examples of how these students' engagement with, and learning of, mathematics evolved over the four-year period.

Participating Students

Data were collected from students across the four years (2010–2013) of the RoleM project. Students who participated in the study attended schools with low ICSEA (Index of Community Social-Economic Advantage) scores: scores of less than 1000. ICSEA is a statistical model that uses a number of variables related to students' background to provide a school with a score (ACARA 2010). The variables considered in the calculation are socio-economic status, remoteness, percentage of Indigenous, and language background (ACARA 2010). The sample was drawn from three different geographical locations across Queensland: metropolitan,

E. Warren and J. Miller, *Mathematics at the Margins*,
SpringerBriefs in Education, DOI 10.1007/978-981-10-0703-3_4

remote, and very remote locations. All schools were considered to cater for students at risk. The results reported in this chapter are for those students who participated in both the pre- and post-testing conducted each year of the project.

Across the implementation of RoleM from 2010 to 2013, 1738 students partook in both pre- and post-testing. Table 4.1 presents the number of students who completed pre- and post-testing based on their demographical details (geographical location and ethnicity) and their school year level. With regards to ethnicity, Indigenous students were either Aboriginal and/or Torres Strait Islander. Torres Strait Islander students are from the Torres Strait Islands. They are distinct from Aboriginal people, and are generally referred to separately. Both groups are Indigenous people of Australia. ESL students are commonly students whose first language is not English. Thus many Aboriginal and Torres Strait Islander students can be classified as ESL students. For the purpose of reporting the results of the RoleM project, ESL students were defined as those students who received government support. In Queensland, the support for ESL students is usually given for their first 3–5 years in Australia (Education Queensland (EQ) 2007). There are two separate funding avenues for ESL students; new arrival funding, a one-off payment to provide intensive ESL support; and LNSLN funding (literacy, numeracy and special learning needs) to provide support in mainstream classrooms (EQ 2007). Students with a parent or both parents born in a non-English speaking country can be potentially supported during Foundation—Year 3. On the whole, schools only identify these students as ESL students. Table 4.1 presents the number of participating students by geographical location and ethnicity.

Of students who participated in the project, 62 % were from metropolitan areas, 30 % from remote areas, and 8 % from very remote areas. Australia is a geographical large sparse country whose population is predominantly situated around the coast of the Australian continent, clustering around the major metropolitan cities. Thus, there are very few people who live in very remote locations, and most who do are Indigenous Australians. Thirty-seven percent of students were in the Foundation cohort. Data were collected for this year level for all four years of the project, and the sample size for each year level reduced as the project progressed.

Table 4.1 Number of participating students by demographic details including geographical location, year level, and ethnicity

	Metropolitan			Remote			Very remote			Total
	N-I	I	ESL	N-I	I	ESL	N-I	I	ESL	
F	86	81	152	101	127	1	8	79	0	635
1	62	60	153	67	101	0	5	46	0	494
2	110	45	119	28	60	1	1	7	0	371
3	89	24	94	1	30	0	0	0	0	238
Total	347	210	518	197	318	2	14	132	0	1738

NB: N-I = non-Indigenous; I = Indigenous; ESL = English as a second language

For example, the Foundation cohort comprised students from 2010–2013, Year 1 cohort comprised students from 2011–2013, Year 2 cohort comprised students from 2012–2013, and Year 3 cohort comprised students from 2013 cohort only. Additionally, the highest proportions of non-Indigenous students were from metropolitan areas, and the highest proportions of Indigenous students were from remote areas. ESL students were predominantly from metropolitan areas. Over the four years of the project the number of students in each group was similar (ESL (520), I (660), N-I (558)).

Description and Development of Assessment Tools Used to Measure Students' Numeracy Achievement

Students' numeracy achievement was measured to assess their growth in mathematical understanding across the school year. Four mathematics tests were developed to ascertain students' mathematical learning. The structure of these tests reflected the format of previous Australian national and state tests (Queensland Year 3, 5, and 7 tests, and NAPLAN numeracy tests). Additionally, they were weighted accordingly to the emphasis that is placed on particular mathematical concepts in the Australian Mathematics Curriculum. Existing validated tests informed the construction of the RoleM tests. For example, *Diagnostic Mathematical Tasks* (DMT) (Schleiger and Gough 2001) and *I Can Do Maths* (Doig and de Lemos 2000), *PATMaths* (ACER 2010), and Okamoto and Case (1996) informed the type and style of some questions. Mapping of the concepts utilised in published tests indicated that there were gaps in the mathematical content of these tests, particularly in the domains of geometry and measurement. Thus, the decision was made to develop four original tests with each relating to the first four years of formal schooling, namely, Foundation, Year 1, Year 2, and Year 3. The final mathematics tests consisted of 24 questions, 30 questions, 39 questions, and 26 questions respectively. The language used to pose each question mirrored the style used in the NAPLAN numeracy tests. The National Assessment Program—Literacy and Numeracy (NAPLAN) is an annual assessment for all Australian students in Years 3, 5, 7 and 9. All tests were piloted before administration in the main study and appropriate adjustments made. Figure 4.1 presents one item from each test. The script on the right hand side of Fig. 4.1 was read out loud to all participating students by the test administrator. As the learning of mathematical language in conjunction with representations was a significant component of the project, the language utilised in the verbal instruction was consistent for all students. No modifications were made for the participating ESL students.

The majority of the questions on these tests were multiple choice.

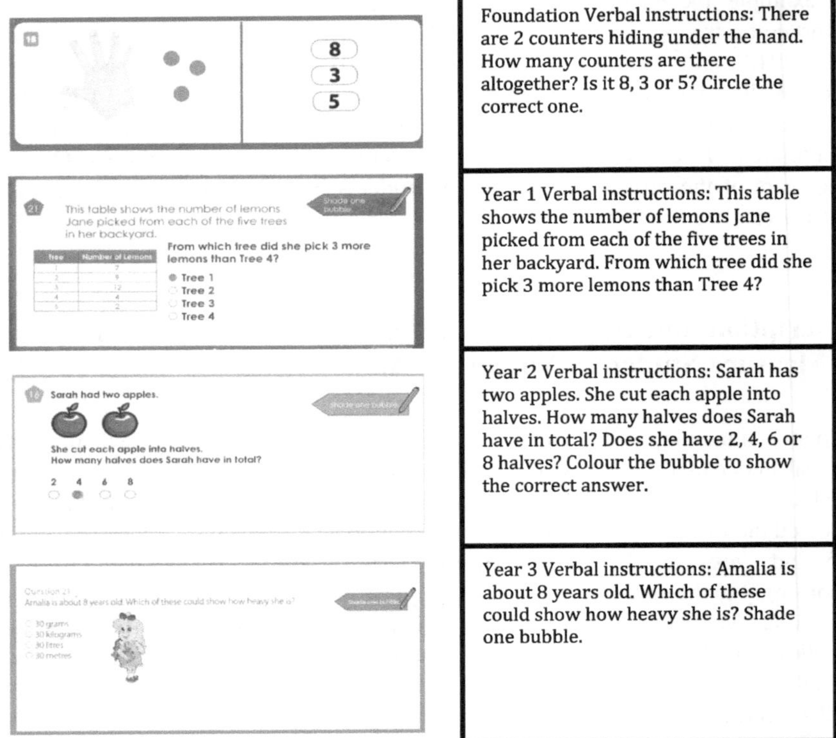

Fig. 4.1 Examples of test questions and verbal instructions

Administration of the Tests

To ensure reliability and consistency across the data set, trained RoleM staff members administered the four tests. Before the administration of each, the test administrators underwent a training program. There was also extensive discussion with regard to the amount of time allowed for the administration of the test. A uniform time of up to 40 min for each test was agreed. Members involved in the administration of the test were assigned the same year level. All tests were read out loud by the administrator, a common practice utilised with students of this age.

The pre-test was administered at the commencement of Term 1 (February/March) and the post-test was administered in Term 4 (October/November). There was approximately a nine-month period between the administrations of the pre- and post-tests.

Analysis of the Test Data

Data were analysed by RoleM staff members. The questions on all four tests were marked as either correct or incorrect, with a correct response being allocated a score of one and an incorrect response being allocated a score of zero. The data were then entered into an Excel spreadsheet where analyses were preformed to ensure accuracy of the data, and that entry errors were eliminated. The data were then transferred to a statistical package (SPSS) where further analyses were conducted.

Benchmark Data

At the end of 2011, Australian Council for Education Research (ACER) were contracted to equate the RoleM tests to the new ACER *Progressive Achievement Test in Mathematics* (PATMaths) norm-referenced scale. At the completion of 2012 all participating students sat the RoleM tests and either the *I Can Do Maths* (ICDM) test, the new PATMaths Plus 1 test or the new PATMaths Plus 2 test, depending on their age. The results of RoleM tests and the new PATMaths test in each grade were compared and an equating shift between each RoleM test and new PATMaths was calculated. This resulted in the development of conversion formulas that allowed all RoleM test results to be anchored on the new PATMaths norm-referenced scale. The results presented in this paper are the converted RoleM test scores for participating students.

These **converted RoleM scores** allow students' results to be mapped onto scales that are relevant to all Australian students; that is, it allows for benchmarking their results against expectations for all Australian students from Foundation to Year 3.

RoleM Student Data: Quantitative Analysis

Results for the Student Cohort 2010–2013

After screening for normality assumptions, paired *t*-tests were conducted for each student cohort to ascertain the significance of the differences between the pre- and post- test scores. Effect sizes were calculated, including Cohen's *d* and eta^2, to measure the strength of the gains made by students participating in the RoleM project.

Cohen's D

Hattie (2009) used Cohen's *d* as a measure to determine what influences student achievement. This measure emphasizes the importance of the magnitude of difference. Hattie (2009) states that, 'when implementing a new program, an effect size of 1.0 would mean that, on average, students receiving the treatment would exceed 84 % of students not receiving the treatment' (p. 8). Cohen (1988) suggests that considering educational outcomes an effect size of d = 0.2 is small, d = 0.5 is medium, and d = 0.8 is large. Furthermore, Hattie claims teachers typically can attain *d* scores between 0.2 to 0.4 per year. Anything, above d = 0.4 is considered educationally significant, and is labeled in the zone of desired effects. The greater the score above 0.4, the more effect the program is having on students' achievement. A more conservative approach is to report eta^2 as a scale to measure student growth for smaller samples.

Eta^2

In this paper eta^2 scores have been calculated for all paired *t*-tests, to report the magnitude of the intervention's effect. The values established by Cohen (1988) for interpreting the effect are: 0.10 = small effect, 0.30 moderate effect, 0.50 large effect.

Results were analyzed focusing on a number of factors that have potential influence on the score. For example, paired *t*-tests were conducted with respect to geographical location, ethnicity, teacher experience, and year participating in the program. The following section reports the results of the paired *t*-tests in relation to these factors across each year level. The average mean score is the difference between the mean for the pre-test (pre mean) and the mean for the post-test (post mean).

Across Years

Table 4.2 presents the results of paired t-tests conducted for all year levels across each year of the project. The highlighted rows represent data from the year in which the RoleM professional learning occurred. For example, the Foundation professional learning was conducted in 2010.

Paired *t*-test results indicate that all students across all years of the project demonstrated statistically significant improvement from their RoleM pre-test score

Table 4.2 Paired Samples *t*-tests for all students split on year participating in the project and year level

	Students	Pre mean	Pre SD	Post mean	Post SD	Avg. mean	t	d	eta^2	p
Foundation										
All	635	58.40	12.01	76.35	13.74	17.94	37.07	1.39	0.68	<0.001
2010	**161**	**56.95**	**11.75**	**76.04**	**15.23**	**19.09**	**17.44**	**1.40**	**0.66**	**<0.001**
2011	155	62.99	11.74	76.74	13.89	13.74	15.32	1.07	0.60	<0.001
2012	177	54.98	11.49	73.94	12.52	18.96	22.13	1.58	0.74	<0.001
2013	142	59.34	11.98	79.27	12.79	19.94	21.25	1.61	0.76	<0.001
Year 1										
All	494	73.80	9.87	86.91	13.85	13.11	26.31	1.09	0.58	<0.001
2011	**191**	**76.70**	**8.52**	**88.17**	**13.72**	**11.47**	**14.91**	**1.00**	**0.54**	**<0.001**
2012	153	70.84	11.25	86.13	15.45	15.29	15.64	1.13	0.62	<0.001
2013	150	73.12	8.96	86.11	12.13	12.99	15.56	1.22	0.62	<0.001
Year 2										
All	371	87.18	8.10	95.75	10.34	8.56	20.79	1.53	0.54	<0.001
2012	**237**	**86.74**	**7.39**	**96.75**	**10.59**	**10.01**	**20.28**	**1.45**	**0.63**	**<0.001**
2013	134	87.96	9.22	93.97	9.65	6.01	8.81	0.41	0.36	<0.001
Year 3										
All (2013)	**238**	**92.82**	**11.04**	**102.43**	**12.00**	**9.61**	**16.64**	**0.83**	**0.54**	**<0.001**

to their RoleM post-test score (p = <0.001). In addition, the majority of students across all years of the project had eta^2 scores that are considered to be very large effects sizes; that is above >0.50 (Cohen 1988). Evidently, Foundation students made the largest gains compared to other year levels. Cohen's d ranged from 1.61–1.07 for Foundation, 1.22–1.00 for Year 1, and 1.45–0.41 for Year 2. These d scores are considered to be in the 'zone of desired effect' as defined by Hattie (2009), and suggest that the program is positively impacting on students' results. Due to the consistency of results for each year level in each year of the project, it is conjectured that the program was sustainable across the years regardless of whether teachers participated in Professional Development days.

Ethnicity

Further paired t-tests were conducted to ascertain if there were consistencies in the results across student ethnicity. For the purposes of reporting the results, students

Table 4.3 Paired samples *t*-tests for all students split on ethnicity and year level

	Students	Pre mean	Pre SD	Post mean	Post SD	Avg. mean	t	d	eta^2	p
Foundation										
N-I	195	63.51	12.07	81.93	13.13	18.42	21.60	1.46	0.71	<0.001
I	287	54.62	10.83	71.35	13.66	16.73	22.50	1.36	0.64	<0.001
ESL	153	59.01	11.90	78.60	11.15	19.59	20.87	1.70	0.74	<0.001
Year 1										
N-I	134	76.50	8.17	92.85	10.88	16.35	19.51	1.70	0.74	<0.001
I	207	69.29	10.48	78.38	12.16	9.10	11.67	0.80	0.40	<0.001
ESL	153	77.53	7.78	93.26	12.13	15.72	18.78	1.54	0.70	<0.001
Year 2										
N-I	139	89.84	7.17	98.66	9.40	8.83	13.79	0.82	0.58	<0.001
I	112	82.07	6.80	89.08	8.43	7.00	8.48	0.64	0.40	<0.001
ESL	120	88.77	8.13	98.60	10.16	9.72	14.39	0.77	0.64	<0.001
Year 3										
N-I	90	94.12	10.22	105.96	11.10	11.84	13.18	1.11	0.66	<0.001
I	54	85.08	10.36	93.61	9.91	8.53	6.08	0.84	0.41	<0.001
ESL	94	96.01	10.33	104.11	11.54	8.11	9.77	0.74	0.51	<0.001

NB: N-I = non-Indigenous; I = Indigenous; ESL = English as a second language

were classified as (a) Indigenous Australian students (I), (b) new immigrant students to Australia whose first language was not English and were classified by the government as ESL students, (c) and all other Australian students who were non-Indigenous Australians (N-I). These were considered to comprise three very broad ethnic groups. Table 4.3 presents the results of this analysis.

Results from paired *t*-test indicate that all ethnicity groups for all year levels made statistically significant gains, and these gains had moderate and large effect sizes. The results further suggest that Indigenous students are now mapping onto the N-I student scores, and this can be seen clearly in the box and whisker plot in Fig. 4.2.

Location

Students' data were clustered by school location. Table 4.4 presents the results for all students across each year level analyzed according to school location (metropolitan, remote, and very remote). Regardless of geographical location, students made statistically significant gains in mathematics.

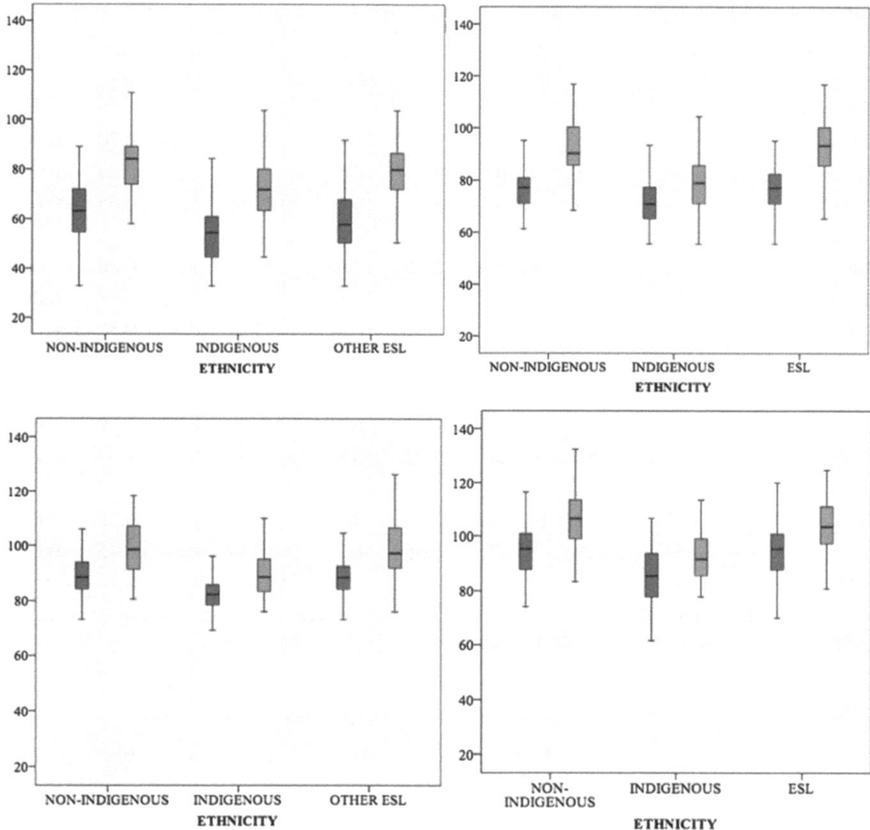

Fig. 4.2 Box and whisker plots for foundation, Year 1, Year 2, and Year 3 split on ethnicity

Teacher Experience

Finally the data were analyzed according to teacher experience. Beginning teachers are those with 0–2 years' experience, experienced teachers having 3–9 years' experience, and very experience teachers have 10 and more years' experience. Table 4.5 presents the results of the paired *t*-tests for students based on teacher experience.

Results indicate that all students made statistically significant gains regardless of teacher experience. One-way analysis of variance (ANOVA) were conducted to determine if there was any significance between teacher experience and the average mean gain for each cohort. The analysis indicates there are no significant differences between groups.

Table 4.4 Paired Samples *t*-tests for Students split on Location and Year Level

	Students	Pre mean	Pre SD	Post mean	Post SD	Avg. mean	t	d	eta^2	p
Foundation										
M	319	59.21	11.66	77.61	12.28	18.40	26.99	1.54	0.70	<0.001
R	229	58.59	12.81	77.10	15.07	18.51	22.31	1.32	0.69	<0.001
VR	87	55.00	11.19	69.76	13.47	14.76	12.65	1.19	0.65	<0.001
Year 1										
M	275	75.61	8.90	90.59	12.76	14.98	22.93	1.36	0.66	<0.001
R	168	72.79	9.60	83.68	13.83	10.89	12.71	0.91	0.49	<0.001
VR	51	67.35	12.42	77.76	12.81	10.40	7.02	0.83	0.50	<0.001
Year 2										
M	274	88.10	7.76	97.34	10.07	9.24	19.98	1.12	0.59	<0.001
R	89	84.71	8.77	90.99	9.83	6.28	7.29	0.42	0.38	<0.001
VR	8	83.16	5.58	94.04	9.23	10.88	2.80	–	–	.026
Year 3										
M	207	94.05	10.63	103.86	11.79	9.81	15.94	0.87	0.55	<0.001
R	31	84.54	10.29	92.87	8.65	8.33	4.94	0.88	0.45	<0.001
VR	0	–	–	–	–	–	–	–	–	–

NB: M = metropolitan; R = remote; V = very remote. Where *d* and eta^2 are not reported, the sample size was too small to conduct these tests

Table 4.5 Paired samples *t*-tests for students split on experience and year level

	Students	Pre mean	Pre SD	Post mean	Post SD	Avg. mean	t	d	Eta^2	p
Foundation										
B	216	56.27	12.78	73.91	13.98	17.65	21.48	1.32	0.68	<0.001
E	181	58.87	12.61	77.78	12.99	18.91	19.69	1.48	0.68	<0.001
VE	238	60.00	10.72	77.47	13.84	17.47	22.96	1.41	0.69	<0.001
Year 1										
B	273	73.92	9.50	85.84	13.37	11.92	17.84	1.03	0.54	<0.001
E	111	73.34	11.19	86.73	13.85	13.39	14.60	1.06	0.66	<0.001
VE	110	73.95	9.43	89.74	14.71	15.79	13.73	1.28	0.63	<0.001
Year 2										
B	200	86.74	8.00	94.52	10.15	7.78	15.13	0.85	0.53	<0.001
E	78	87.38	8.69	96.78	11.15	9.40	8.50	0.94	0.48	<0.001
VE	93	87.97	7.85	97.53	9.76	9.56	12.32	1.08	0.62	<0.001
Year 3										
B	119	92.01	10.22	102.56	11.97	10.56	13.33	0.95	0.60	<0.001
E	56	90.18	11.89	99.01	11.34	8.84	7.76	0.76	0.52	<0.001
VE	63	96.69	10.92	105.21	12.05	8.53	6.98	0.74	0.44	<0.001

NB: B = beginner; E = experiences; VE = very experienced

Exploring Students' Gains

Test items were further analyzed to determine which math concepts students made the greatest gains on. Paired *t*-tests were conducted for each item, and eta^2 were calculated and ranked. These ranked items were then aligned with the concepts delineated in Sarama and Clements' (2009) early years mathematic development progression trajectories. Table 4.6 presents the five mathematics concepts that Foundation students made the greatest gains on, together with example questions from the test. QR codes link each item to videos that provide examples of learning activities students engaged into support the development of this concept. The first column identifies the particular concept in the mathematic development progression trajectory (Sarama and Clements 2009) each item aligned with. Tables 4.7, 4.8 and 4.9 present the results for Year 1, Year 2, and Year 3 respectively.

Foundation

Foundation students made the greatest gains in the area of counting. They demonstrated an ability to produce and count numbers between 0 and 20. It is also evident that Foundation students were able to work across different representations of number including language, symbols, and materials.

Table 4.6 Foundation concepts, test items, and examples of learning activities

Foundation	Example test question	Learning activity	QR code
Counter and producer of numbers 10 and larger	Stick 10 yellow stickers in the large box. Write the number in the green circle	Linking the three representations of number	
Counter to 10	Circle all the numbers that you can see	Beginning to Count	
Counter and producer of numbers 10 and larger	Point to the number in the green circle (11). Stick that number of yellow stickers in the large box	Linking and ordering a number of objects to digits	
Counter and producer of numbers 10 and larger	Write the number that comes after 4, 9, 17	Counting on and counting back using a number ladder/track	
Composer to 10	Stick 7 stickers in the large box. Write the number 7 in the green circle	Linking and ordering a number of objects to digits	

Table 4.7 Year 1 concepts, test items, and examples of learning activities

Year 1	Example test question	Learning activity	QR code
Counter to 100	48 … 54 / 48 … 54	Sequencing numbers on a number track	
Counter to 100	26 … 34 / 26 … 34	Counting on and counting back using a number track	
Counter to 100	26 27 28 29 37 / 26 27 28 29 37	Exploring the structure of the 100 board	
Problem solver	There were 21 frogs swimming in the pond and 2 jumped out. How many frogs are left in the pond? / There were 21 frogs swimming in the pond and 2 jumped out. How many frogs are left in the pond?	Representing and solving simple take-away and missing addend word problems	
Time—length unit relater	What time is shown on this clock? 7:00 7:30 6:30 8:30 / What time is shown on this clock? 7:00 7:30 6:30 8:30	Reading times to the hour and half-hour	
Counter on using patterns	What is the missing number in this number pattern? 23, 25, 27, 29, __ 33, 35 / What is the missing number in this number pattern? 23, 25, 27, 29, __ 33, 35	Skip counting by 2s and 5s	

Table 4.8 Year 2 concepts, test items, and examples of learning activities

Year 2	Example test question	Learning activity	QR code
Data—reads and represents data	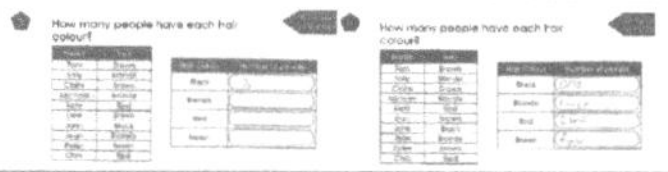	Gathering and interpreting data	
Area as an array	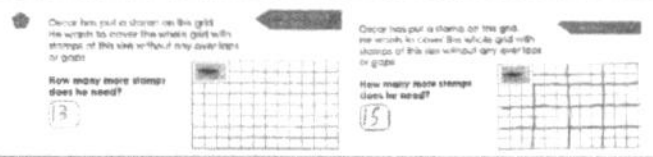	Measuring area	
Subtraction word problems	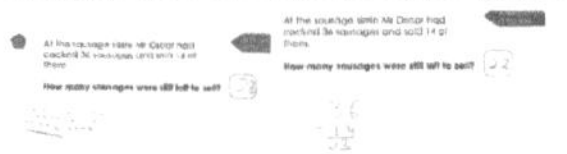	Unpacking subtraction word problems	
Representing multiplication as groups of	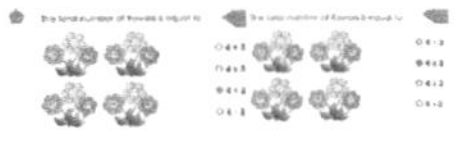	Recognizing groups and arrays	
Missing addend addition problem		Creating, writing, interpreting simple addition and subtraction word problems	

While these test items were the questions that students made the greatest gains on, they were not the questions students performed best at in the post-test. The following list identifies the top five best-answered questions with the percentage of students who answered them correctly:

- identifying and translating a missing piece of a puzzle (77 %)
- continuing an ABABAB repeating pattern (74 %)
- completing a repeating pattern (73 %)
- counter and producer to 10 (71 %)
- connecting number names, numerals, and quantities (67 %).

Year 1

Year 1 students' gains were again in the area of number with more complex conceptual understanding of number being attained. For example, they showed an awareness of 'counting on' using number patterns.

The following list identifies the top five best-answered questions with the percentage of students who answered them correctly:

- adding two single digit numbers together in separate 10 s frames (74.3 %)
- classifying, counting and recording data (71.5 %)
- identifying how many halves in 4 wholes (71.3 %)

Table 4.9 Year 3, Test items, and examples of learning activities

Year 3	Example test question	Learning activity	QR code
Division into groups of 10 (120/10)—or repeated aggregation	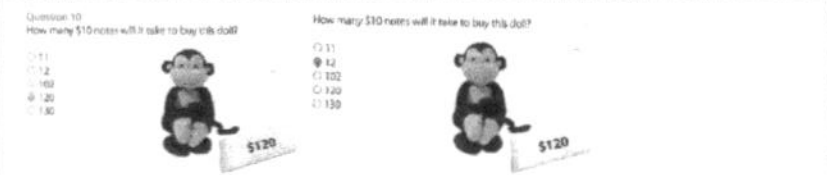	Decoding division and multiplication word problems (equal groups)	
Multiplication representations and structures (×2)	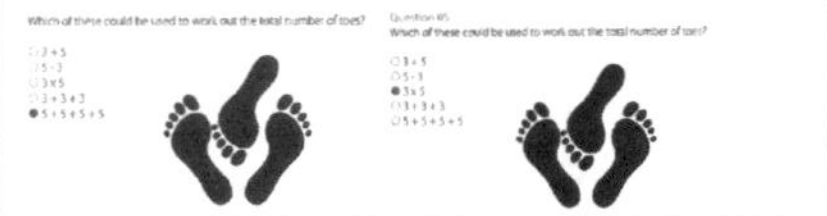	Exploring multiplication as equal groups	
Multiplication representations and structures (×2)	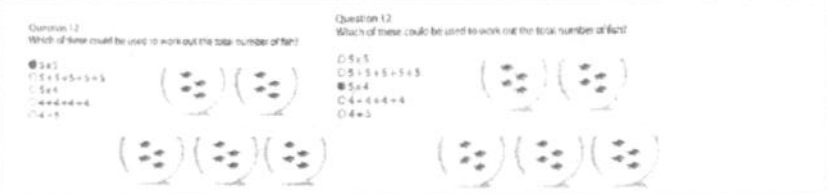	Exploring multiplication as equal groups	
Correct money—counting in 50s 20s 10s	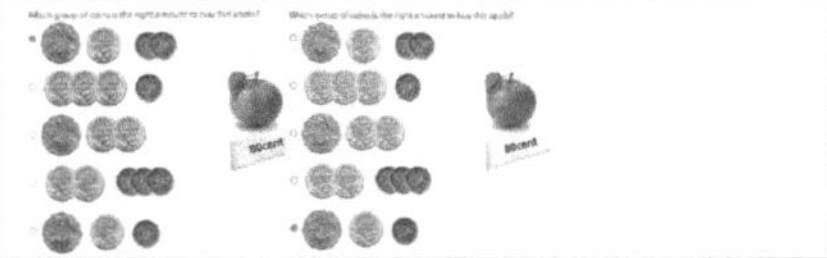	Representing money values in multiple ways	
Shape and location and movement—mental mover	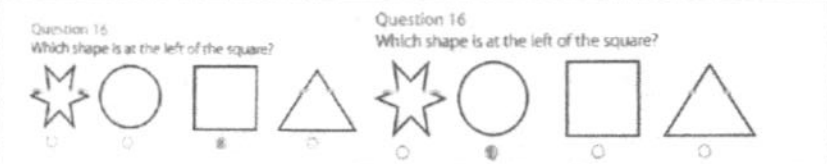 Question 16 Which shape is at the left of the square? Question 16 Which shape is at the left of the square?	Creating and interpreting maps using Bee-bots	
Place value 3 digit numbers	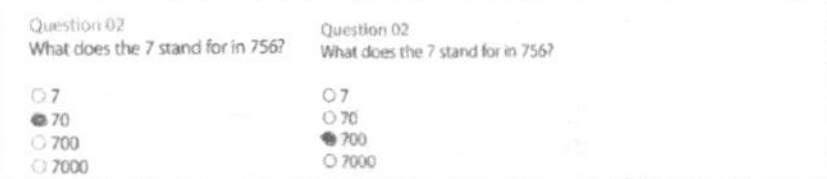 Question 02 What does the 7 stand for in 756? ○ 7 ● 70 ○ 700 ○ 7000 Question 02 What does the 7 stand for in 756? ○ 7 ○ 70 ● 700 ○ 7000	Reading and writing four digit numbers	

- single digit addition word problems using materials (71.3 %)
- identifying the representation of 62 in tens frames (69.4 %)
- calculating the missing addend for single digit problems (63.2 %).

Year 2

The greatest gains made by Year 2 students were spread across mathematical concepts. The concepts included reading and recording data, measurement with respect to area as arrays, multiplication, and word problems.

The following list identifies the top five best-answered questions with the percentage of students who answered correctly:

- identifying parts of a repeating pattern ABBBCC (82 %)
- missing addend—represented as a length model (67.1 %)
- using array model to solve multiplication (65.2 %)
- identifying repeats in a pattern ABABAB (64.7 %)
- solving two digit missing addend problem traditional algorithm (61.9 %).

Year 3

Year 3 students made most gains in the area of number. The areas they made gains in were more sophisticated and more difficult (e.g., division into groups of 10) as compared with previous years.

The following list identifies the top five best-answered questions with the percentage of students who answered them correctly:

- identifying the correct day of the week using a calendar (89.9 %)
- direction and location (89.1 %)
- symmetry (83.2 %)
- identifying a 2-digit number with representations of tens and fives (82.8 %)
- place value (82.4 %).

Hypothesized Trajectory of Mathematics Concepts

Considering the above results, a map of the concepts students moved through as they participated in the pre- and post- testing was developed. Each item was selected according to the percentage frequency of students who successfully answered the item in the posttest. If the percentage frequency was larger than or

Table 4.10 Trajectory of mathematical concepts demonstrated by students across the four years of RoleM

Mathematical concepts					
Areas of mathematics	Probability			Identifying if an event was likely, unlikely, certain, or impossible	
	Statistics		Count and record data	Read and interpret data (bar graphs and tallies)	Interpret and infers data
	Geometry	Identify and classify 2D shape (rectangles) Classify 3D shape (curved surfaces) Translation of shape	Movement and location	Identifying a flip (reflection) Reading a map/plan	Symmetry of 2D shapes Identifies triangles Identifies faces of 3D shapes Reading angles and identifying largest angle Location and direction (multiple directions)
	Measurement		Represent both analogue and digital time Identify half hour time	Identify the correct length of an object	Estimates weight (kg) Reading a calendar
	Patterns and algebra	Continue a repeating pattern Complete a repeating pattern Create a repeating pattern	Identify unit of repeat	Completing a complex repeating pattern	Completes simple number patterns

(continued)

Table 4.10 (continued)

Mathematical concepts					
	Number	Counter and producer to 10+	Identifying missing numbers in a number sequence (2 digit) Identify a two digit number is composed of tens and ones Identifies money representations Solves money problems involving dollars and cents Single-digit addition from representation to symbol Addition word problems Missing addend problems Identifies how many halves in a whole for more than one object	Solves 2-digit missing addend problems Solve addition word problems (a + b + c) Multiplication represented as arrays Using an array model to solve multiplication	Place Value of three digit numbers Solves three digit subtraction Solves subtraction word problems (simple and comparison models) Solves division as sharing problems Identifies 4 digit representations and links to language
		Foundation	Yr1	Yr2	Yr3

equal to 50 % then the item was selected for inclusion in the map, and categorized according to the mathematics concepts tested. Finally, the concepts were plotted against existing curriculum and learning trajectories (e.g., Sarama and Clements 2009). Table 4.10 presents a trajectory of the mathematical concepts demonstrated by students across the four years of RoleM. It appears that these students were performing similarly to other trajectories presented in early years mathematics literature (e.g., Sarama and Clements 2009) and demonstrating an understanding of age appropriate mathematical concepts as delineated in Australian curriculum documents.

Student Change

Teachers reported as they implemented RoleM in their classrooms they observed changes in students' learning. Chapter 3 presents teachers' journey as they participated in professional learning in the RoleM project. As part of the data collection, teachers not only reported on changes they themselves experienced but also the changes they observed from their students as these students engaged with the learning activities implemented in the classroom. Table 3.12 presents subthemes and trends for teachers in order of importance. The dimensions related to student were:

- Engagement and enjoyment (84 %)
- Student cognitive challenges (language and prior knowledge) (81 %)
- Making connections (75 %)

These subthemes were evident across all three interviews conducted each year in the RoleM project and also there was agreement of these themes across geographical location.

As students moved through the RoleM project it was evident that they experienced changes in the learning of mathematics. The most frequent response given by teachers was that they observed that their students had increased their engagement and enjoyment in learning mathematics (84 %). *There were a few of them that weren't very interested in maths originally, but because of all of the hand on activities, they're keen to do it and they're excited when we're doing our maths rotations. It's good that they actually want to do maths now* (1602, 2011). *They are really engaged. When we were working with the hundreds grid every single child in the class was involved and totally engaged and there was great dialogue going on between the kids* (907, 2011). Figure 4.3 presents students engaging in RoleM learning experiences.

As students became more engaged teachers reported that they were able to observe students demonstrating their mathematical learning in class. *They know more. They have an improved ability to talk about how they come up with ideas. They have a willingness to say I did that because of this and this* (802, 2010). *It was interesting to see the students understood place value. One of the questions was say*

Fig. 4.3 Students engaging in RoleM learning experiences

the number was 533; *students could demonstrate that is was* 500 + 30 + 3 (207, 2012). Figure 4.4 presents students demonstrating their mathematical learning.

Teachers reported that they believed one challenge for students was their mathematical language and prior knowledge in mathematics. As students engaged in the RoleM activities they were able to demonstrate the different ways of learning that assisted them to overcome these challenges. *The concrete materials that we use in RoleM, they help my children because, if we are talking about a concept and they are finding it abstract, once I bring out the materials and I have them doing things with me, they generally will pick it up 'oh that's what she means'. It helps break down that language barrier* (305, 2013). *I have an ESL boy with very limited understanding of any concepts. I did the multiplication activity with the groups of counters on with the plates, and I honestly did not believe that this child would be able to understand the concept of 'there is three plates and there are four buttons on each, how many altogether?'. He understood it and he can now automatically do* '3 *times* 4 *is* 12'. *He has got to that stage of automatically knowing his tables from using the plates. Our HOC videotaped him and then showed him to the staff of how a child that I had a lower expectation of, or a different expectation for, actually achieved, or did better than the children I had a really higher expectation of* (909, 2012).

Figure 4.5 presents students demonstrating the different ways they learn mathematics.

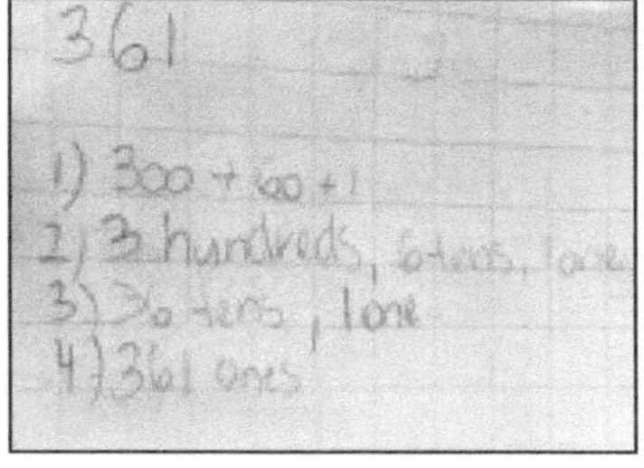

Fig. 4.4 Students demonstrating their learning

Fig. 4.5 Students demonstrating the different ways they learn mathematics

Teachers also reported changes in students making connections between mathematics and the real-world. *The kids can relate to mathematics and they can see a connection between the real-world* (807, 2012). *They are getting good at doing it in similar contexts. So the use of more and less, I am hearing it more in their own play. They are actually discussing those things, whether there are more girls or more whether there are more boys…they are keying into it before i even start that floor session time. A lot of games—they are using their counting and they will say 'you need 1 more piece' or 'you've got too many pieces, you need to have less". They are actually using the formal language rather than how they would normally say it* (102, 2010). *In our general day to day way that we do things…the kids are making connections with the outside world of what we're doing, so they will start looking at shapes and talking about sides and all the things we have been talking about, but not in that context. They are developing it in their own play that they are doing. In negotiated play with themselves a lot of connections are those, that language we are using coming out* (108, 2010). Figure 4.6 students making connections with mathematics and the real-world.

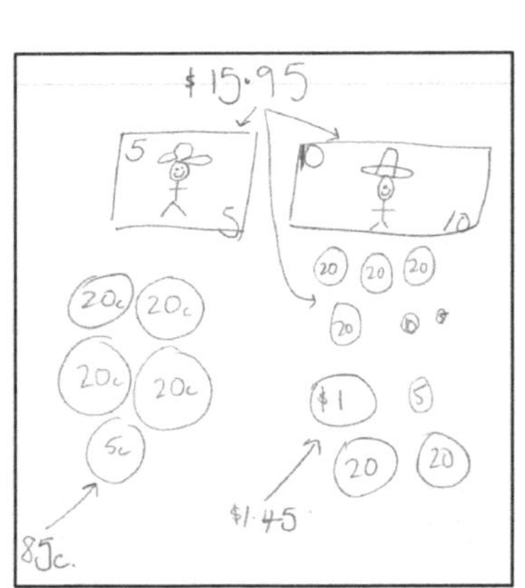

"I'm swinging under the of the monkey bars"

Fig. 4.6 Students making connections between mathematics and the real-world

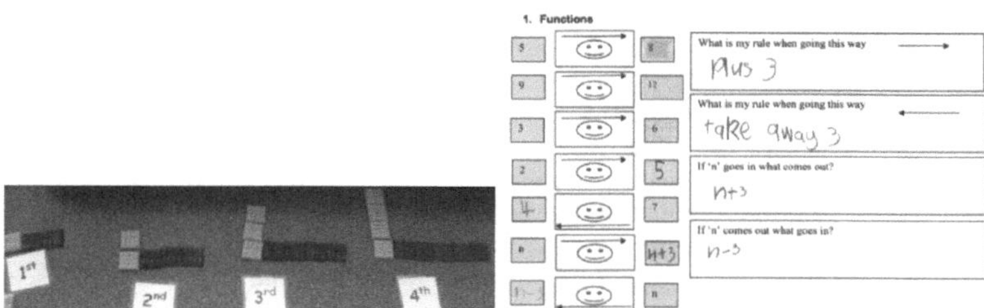

Fig. 4.7 Students demonstrating higher levels of mathematics

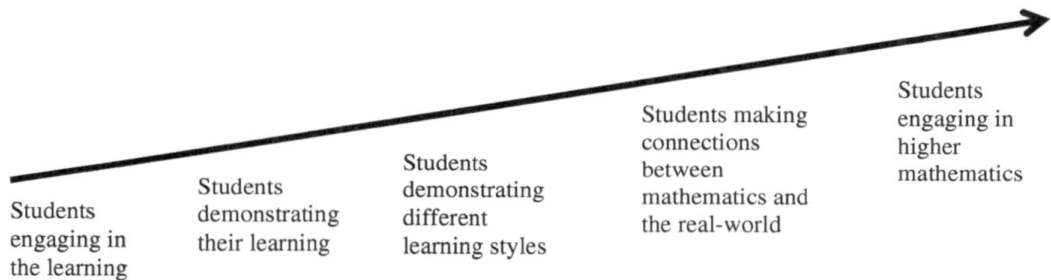

Fig. 4.8 Hypothesized student-learning trajectory

Finally, students were reported to engage in higher levels of mathematics. Not only was this demonstrated in the post-test results in RoleM across all year levels but teachers also reported observations of this occurring in class. *It was amazing to see the students creating their own growing patterns and engaging in early algebraic thinking. They were able to predict what the pattern would look like and how you would make it at positions beyond the patterns given* (316, 2012). *Students are now demonstrating different problem solving strategies and I am able to use the extension activities to support my students* (108, 2013). Figure 4.7 demonstrates students engaging in higher levels of mathematics.

From these observations and analysis of the teacher and student data a hypothesized student-learning trajectory emerged (see Fig. 4.8). It appears that the students move through five stages.

In summary, RoleM had positive impacts on students' mathematical learning. All students across all year levels, regardless of geographical location, ethnicity, or their teachers' experience, had statistically and educationally significant gains in mathematics. The trajectory of mathematical concepts presented in this chapter is a snapshot of students' learning measured by the RoleM post-test. The student-learning trajectory presented in Fig. 4.8 will be discussed further in Chap. 6.

References

Australian Council for Educational Research. (2010). *Progressive Achievement Tests in Mathematics*. Canberra, ACT: ACER.

Australian Curriculum, Assessment and Reporting Authority (ACARA). (2010). *Calculating ICSEA values*. Retrieved from http://www.acara.edu.au/verve/_resources/Calculating_ICSEA_Values.pdf

Cohen, J. W. (1988). *Statistical Power Analysis for the Behavioural Sciences*. Hillsdale, NJ: Lawrence Erlbaum Associates.

Doig, B., & de Lemos, M. (2000). *I can do Maths*. Melbourne, VIC: ACER Press.

Education Queensland (EQ). (2007). *ESL Handbook for Schools*. Retrieved from: http://education.qld.gov.au/studentservices/inclusive/cultural/esl/docs/schoolseslhandbook2006–07.pdf

Hattie, J. A. (2009). *Visible Learning: A Synthesis of 800 + Meta-Analyses on Achievement*. Abingdon: Routledge.

Okamoto, Y., & Case, R. (1996). Exploring the microstructure of children's central conceptual structures in the domain of number. *Monographs of the Society of Research in Child Development, 61*, 27–58.

Sarama, J., & Clements, D. H. (2009). *Early Childhood Mathematics Education Research: Learning Trajectories for Young Children*. London: Routledge.

Schleiger, H., & Gough, J. (2001). *Diagnostic Mathematical Tasks*. Sydney: UNSW Press.

Chapter 5
Crossing the Divide

Abstract The demographics of these teachers closely align with international and national trends for teachers in marginalized contexts. The majority of these teachers were inexperienced and under qualified (Borman and Kimball in Elementary Sch J 106(1):3–20, 2013). This problem was exacerbated as the geographical location became more remote. The more remote the location the higher the percentage of inexperienced teachers, and the more under qualified and under-confident they felt. This chapter discusses the teacher and student data with regard to the literature, and identifies the particular dimensions that assisted these teachers and students to succeed in teaching and learning Western mathematics.

Participating Teachers

At times the participating teachers seemed overwhelmed by the contextual issues they faced, and these issues were amplified for inexperienced teachers in remote and very remote locations. The participating schools were under-resourced (Kent 2004). There were high levels of absenteeism and behavioral problems, and students' lack of Australian Standard English impacted on teachers' perceived ability to effectively teach in these contexts (Allard and Santoro 2004). In addition, the participating inexperienced teachers were under confident in implementing a mathematics program in their classes (Hewitson 2007). But at odds with the literature is that all these teachers were committed to helping their students learn. Many recognized that their practices needed to change, but were unsure just what this entailed (Jorgensen et al. 2010).

As the year progressed 95 % of teachers shared that the three main areas they made the greatest gains in were: their confidence to teach mathematics to these marginalized students; their ability to plan and organize the teaching of mathematics in their classroom; and, their ability to differentiate the learning (see Table 3.12). On a closer examination of the data, teachers' gains in self-confidence impregnated

E. Warren and J. Miller, *Mathematics at the Margins*,
SpringerBriefs in Education, DOI 10.1007/978-981-10-0703-3_5

Table 5.1 Planning and organizational sub-themes across the year

Interview 1	Interview 2	Interview 3
It [RoleM] has made me think about the way I structure [organize] Maths in my classroom (207, 2012)	Easier to work in small groups… more specific in what I am looking at… getting a chance to talk with kids about how they think (102, 2010)	Have more activities out, and they are still learning that concept but they are doing it independently as well (1701, 2013)

all areas of their teaching of mathematics (e.g., planning and organizing, ability to differentiate, knowledge of mathematics), and the main impetus for this change was their engagement in the RoleM professional development and using the RoleM resources in their classroom. In each section representative quotes are presented to give insights into the changes that occurred across the year. Member checks from other researchers gauged the authenticity and robustness of these quotes as representations of these changes.

The planning and organizational gains that teachers made were closely aligned with their beliefs about their instructional efficacy, their beliefs in their ability to effectively teach mathematics to their students. The relationship between teachers' instructional efficacy and socio-economic background is a double-edged sword (e.g. Bandura 1991). Teachers often enter these marginalized contexts with a mistaken belief that these students are incapable of exhibiting age appropriate learning outcomes (Jorgensen et al. 2010). They also doubt their own ability to teach mathematics to these marginalized students (White and Reid 2008). Thus, a teacher's instructional self-efficacy is heavily influenced by beliefs about students' capabilities. An examination of representative quotes for the planning and organizational sub-theme, presented in Table 5.1 give greater insights into this relationship and the shifts that occurred to these constructs as the year progressed.

As the year progressed, teachers were directly linking their increased beliefs about their ability to teach mathematics to their increased beliefs about their ability to gauge what students know and can do. While this relationship is acknowledged in the literature (Bandura 1992), what is lacking is insight into the strategies that help these shifts to occur.

Provision of Efficacious Resources

The strategy that appeared to have the most influence on this shift was the RoleM resources developed by the team and given to teachers to utilize in their classroom.

> I can use them in all different ways, for group lessons as well as independently, turning them into games (303, 2011). They help students gain an understanding of mathematics (501, 2010).

The RoleM resources significantly contributed to teachers' ability to differentiate the learning for the wide range of student abilities. The different ways that teachers quickly changed the learning activities to suit students' capabilities was consistency referred to across the three interviews (see Table 3.7).

> When the five frames were introduced [at the PL] moving into the 10 frame concept and then moving into the 20 frames. That was of particular interest to me because I have got lots of kids that are at all of those levels in my class. It was really good for me to see how I can integrate all that across the learning (301, 2011). [The resources helped me to] learn a variety of different ways to explain and demonstrate concepts (207, 2012). Those activities allow students to play with them. Students get excited about the maths. Their understanding is better (201, 2010).

While knowledge of tasks as instructional tools is acknowledged as a fundament of pedagogical knowledge (Krauss et al. 2008), its link to teachers' ability to cater for the diversity in classroom contexts is underexplored. The results of this study begin to tease out this relationship. Although there was a dimension of open-endedness to the learning activities—a feature seen as a key dimension for catering for diversity (Boaler and Staples 2008)—the dimension that these teachers found most useful was the explicit layering of concept development and the continual deepening and extension of concepts as they moved through the learning activities across the year. This deepening occurred by using a variety of representations, a variety of language structures, and targeting different learning styles.

The debate over regard to how to effectively integrate representations into classroom learning remains ongoing (Cooper and Warren 2008). The results of this research confirm some of our theories hypothesized from the results of our Early Algebra Teaching Project (Warren and Cooper 2009). The next section presents each of our conjectures related to these theories, together with explicit examples taken from the learning activities. Figure 5.1 illustrates conjectures 1 and 2.

1. *Effective initial representations show the underlying structure of mathematical ideas and easily extend to new concepts and new applications.*
2. *In an effective sequence representations develop in three ways (i) from concrete to diagrams to symbols (ii) increased coverage where later representations*

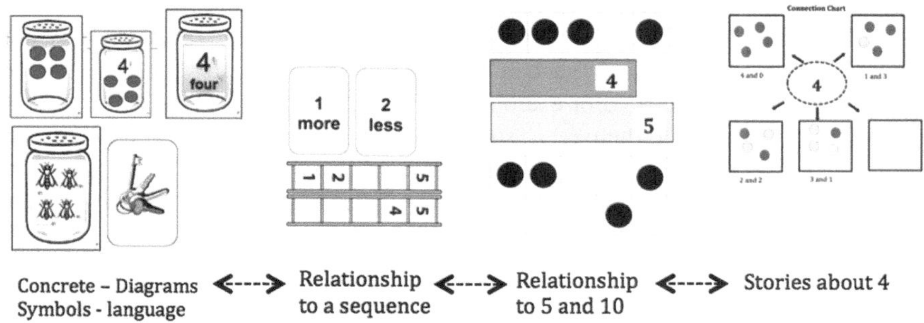

Fig. 5.1 The number 4

compensate for the limitations of early representations; and (iii) connected to reality, always relating the form of representation to real world instances.

3. *Abstraction of a concept occurs across models and representations that follow a structured sequence.*
4. *Where possible later representations should be subsets of earlier representations, gradually leading to abstraction.*

With regard to culture and contexts, the ITAs and classroom teachers drove changes to the representations utilized at each site. For example, in one of the Island communities, kangaroos are non-existent. Thus the pictures of objects to count were changed to turtles. Another example is the use of the flies in the jars. One community suggested we change flies to spiders as these were more prevalent in their environment.

In summary, we are suggesting that these key aspects of the RoleM resources best supported the evidenced changes in these teachers, and subsequently their students. This claim is supported by the student pre and post-test data presented in Chap. 4 where it is clearly shown that the program is positively impacting on students' results. These are also the aspects of the program that teachers found most helpful in supporting their own change (see Chap. 3).

Emphasis on the Language of Mathematics

The emphasis on the language of mathematics in the resources and the professional learning resulted in teachers feeling that they could now effectively communicate mathematical concepts to these students. This also assisted these teachers to shift from their increased beliefs about their ability to teach mathematics to their increased beliefs about their ability to gauge what students know and can do.

> I put a lot more emphasis on the language of maths now than ever before. I try to integrate the language where possible (801, 2010). … a lot more different terms. Instead of using the same language all the time, like more or less, using terms so the kids are getting more variety (802, 2011). They in turn catch on and use it as well (1504, 2013).

We contend that the success of this particular approach to language development in mathematics (see Fig. 2.2) requires embedding language to related representations and real world situations that can be acted out kinesthetically and/or visualized (Warren and Cooper 2009). For example, in Fig. 5.1: at the concrete stage students placed objects familiar to them (e.g., sticks, stones, buttons) in their jars to represent each number; at the relationship and sequence stage large ladders were placed on the floor so that students could 'walk the numbers' as they counted forward and backwards; and, at the relationship stage students represented their stories using tactile objects such as counters, small plastic fish, and small plastic kangaroos. Attention to this ensures that such interactions focus on mathematical development rather than become an exercise in linguistics (McDonald et al. 2011). The transfer of this 'new language' to other contexts is evidenced by the following teacher quote:

> It is just a matter of using it in everyday life around the school, in the classroom, in the playground. From me using mathematical language [with appropriate visual representations] they [students] in turn catch on and use it as well (1504, 2013).

Consequential Gains in MCK and PCK

Finally, teachers made substantive gains in their mathematics content knowledge (MCK) (71 %) and pedagogical content knowledge (PCK) (70 %) as each year progressed.

> More explicit [MCK] work around the fundamental parts of maths that you need to constantly revise (805, 2011). There are so many more concepts and ways [of teaching the concept] that is what I found (302, 2010).

The PCK predominantly related to their increased ability to help these students effectively communicate their understanding of mathematical constructs.

> I've incorporated language a lot more into when they are playing. Things like, questioning, modeling language, getting students to think about it [language] while they're doing it [activities]... How is it [an object] moving, how is it rolling? (402, 2010)

Teachers' focus on language and communication is not surprising, as many students in these marginalized contexts in Australia do not have Standard Australian English (SAE). As all teachers enter these contexts the first notable characteristic they observe is their inability to effectively communicate with their students. Meaney et al. (2008) claim that in order to teach and assess using SAE students need to possess an adequate linguistic repertoire. The results of our research suggest that with regard to mathematics this claim is too narrow, as it puts the focus solely on the linguistic dimension. Thus, we claim teaching mathematics to these students requires a rich repertoire of purposely-selected representations in conjunction with a rich repertoire of how to communicate about these representations, including language, gestures and movement. It is when students engage with this discourse that they are deemed to be communicating mathematically (Moyer 2000). As Jones et al. (1995) argue, acknowledging and targeting just the cultural and linguistic factors that influence the learning of Western mathematics in these classroom is not enough.

The results of this research confirm that MCK and PCK are related constructs (Baumert et al. 2010). An examination of the trends in the interview data suggest that while these developed simultaneously (see Tables 3.7 and 3.8), teachers' improved understanding of teaching language predominated across all three interviews. These trends contradict Kleickmann et al.'s (2013) finding that the gains made once teachers are teaching are predominantly for PCK. The gains made in MCK by these teachers may simply reflect the fact that teachers in marginalized contexts tend to have less subject matter knowledge (Hill et al. 2005) than mainstream teachers. There is room to learn. We suggest that the RoleM professional learning model, unlike traditional Professional Development, positively impacted on this development (Darling-Hammond and Richardson 2009). The conjectured relationship between PCK, MCK and PD are further discussed in Chap. 6.

Participating Students

Results indicate that engagement in the RoleM project had positive outcomes for all students. National and international trends indicate that students from marginalized contexts perform below standard in mathematics (e.g., OECD, NAPLAN, TIMSS), however, results from our study indicate that marginalized students can have success in mathematics. The Cohen's *d* scores as reported in Chap. 4, demonstrate that students were making large shifts in their mathematical knowledge from the pre-test to the post-test. These gains are not solely developmental; the implementation of the mathematics program significantly contributed to these gains (Hattie 2009). Hattie (2009) suggests that Cohen's *d* scores above 0.4 are attributed to the program implemented by the teachers or school. Seventy percent of the Cohen's *d* were above 1.00 and a remaining 21 % were above 0.6 (see Table 4.2).

While these students may have entered school with little prior knowledge of Western mathematics, they were successfully engaging with mathematics concepts by the end of their first year of school. The greatest gains made between the pre- and post- test scores were by the Foundation students. This is in contrast to a previous study in Queensland with large cohorts of young students in Queensland (involving 132 teachers and their 1831 students). This study found that many Year 1 students made negative progress in their understanding of basic numeracy (Thorpe et al. 2004). It is conjectured that this occurred due to (a) these early years teachers lack of MCK and PCK, and (b) the emphasis that they tend to place of literacy at the expense of mathematics. This also contradicts the findings of a large international study conducted by Denton and West (2002). They found that students from low-income families gained little understanding of mathematics in their first year of school. The results from the first four years of RoleM clearly show that all Foundation students are capable learners in formal schooling contexts when given the opportunity and an approach that is conducive to engaging them in the learning process.

Teachers' experience did not have a significant impact on students' gains in mathematics achievement from Foundation to Year 3 (see Table 4.5). Hill et al. (2005) claim that even though teachers experience does not impact on Year 1 students achievement, it does begin to a have marginally significant, positive impact by third grade. Our findings do not align with this trend. This could be due to two reasons. First, the magnitude of the gains these students made was so great that it is impossible to gauge any marginal differences that teachers' experience has on students' achievement (Hattie 2009). Second, the measure of experience is related to the number of years that these teachers have spent in these contexts. Even after a number of years in these contexts many of these teachers are still struggling, with many adopting traditional (inappropriate) pedagogies in order to survive.

Results from RoleM are in contrast to the trends found in national and international testing with respect to geographical location. Regardless of where students were situated in Queensland (metropolitan, remote or very remote), they made significant gains in their mathematics scores. Data from the 2014 NAPLAN tests report that the national average for Australian students was 401.8 (SD = 73) (ACARA 2014). Queensland students from metropolitan areas were 3.6 points below the national average, remote areas were 33 points below the national average, and very remote locations were 67.6 points below national average (ACARA 2014). By contrast, we found that there was no significant difference between students' scores irrespective of geographical difference.

Finally, all students regardless of ethnicity successfully engaged with the test that reflected the structure and language utilized in national and international numeracy tests. This aligns and further extends our past research with Foundation students (Miller and Warren 2014). Unlike Shaftel et al.'s study (2006), it appears that the linguistic features of the RoleM tests did not impact on the ESL students' ability to perform (Miller and Warren 2014). Issues for ESL students pertaining to syntactic, sematic and pragmatic features of mathematics language (Spanos et al. 1988), appeared to be no different to that of the mainstream students tested. Data from this study indicate that ESL students outperformed Indigenous and non-Indigenous students as evidenced by post-test scores for Years 1, 2, and 3 (see Table 4.3). In addition, there was no need to simplify the language used in the test to have a positive impact on students' performance, unlike past studies (e.g., Abedi and Lord 2001). The results also support the claim that the language of instruction used by teachers in the mathematics classroom does not appear to make a quantitative difference within multi-lingual settings (Valle et al. 2013). We contend that what makes a difference is the way in which students' mathematical language is developed.

The use and development of mathematical language needs to be contextualized for ESL students living in marginalized contexts, and these contexts should include the use of a wide variety of mathematical representations. When using an ESL approach (an approach that predominantly involves mapping from one language to another language) to teaching mathematics the interactions can become a linguistic exercise rather than an exercise in developing an understanding of mathematics (Moschkovich 2007; McDonald et al. 2011). Separating the language from its representations can lead to misconceptions in mathematics. It could be argued that acknowledgement of the cultural and linguistic factors that affect the learning of Western mathematics are a minimalist type of action that has been ineffective. Such a focus in the mathematics classroom can inappropriately result in a focus on the use of the English language rather than on the acquisition of mathematical concepts (Celedón-Pattichis and Turner 2012; Howard 1997).

Students increased their knowledge of, and confidence in using, mathematical language while participating in RoleM. They exhibited an increased use of mathematical language within and beyond the classroom, and an increased confidence in discussing mathematics with their teacher and peers. It appears that the learning activities students engaged with provided opportunities for them to develop their mathematical language. For many students from these contexts their home language, at times, does not reflect the Western mathematical language used in school contexts (Cairney 2003; Moschkovich 2007).

> I sat down with Sally and she just knew her stuff [mathematical language] and if I had to say it, they (she) would be my lower ones and when it came to the language test they just aced it! (302, 2010). Just when the kids are in the activities, they're starting to mimic my language so you can see they're – like in the beginning of the program you'd ask a question and you'd get a word [single-word answer] answer but now it'd be more well 3 comes after 2 but it is also before 4. So they are really picking up the language. The same language we see [in] the maths games, then they go off and use it in their own time, just in everyday conversation. (802, 2010). One of my little boys who doesn't really speak often really enjoyed having a go and he was being very vocal and explaining things in great depth which was really good to see (1013, 2013).

As the year progressed there were shifts in students' self-confidence and willingness to engage with and discuss mathematics in class. Students in marginalized settings are often under-confident with Western mathematics (Nasir 2011). This increase in confidence is a priority in national and international agendas for mathematics learning (ACARA 2009; NCTM 2000).

> The main thing I have observed is their confidence bloom, it has been amazing just to see them always willing to have a go and not be afraid if they get a question wrong or right. Especially during the final testing, at the beginning of the year I had students crying during it. This time round it was amazing to see that they all had a go and tried their best (1009, 2011).

We suggest that as students' confidence increase so does their willingness to engage with higher levels of mathematics. These gains in turn lead to teachers providing more challenging tasks, tasks that require more cognitive effort to complete (Linnenbrink and Pintrich 2003).

> Their [Students] expectations of themselves have increased as well and there is a lot more conversation around mathematics and why things are what they are. Definitely I think that their expectations of themselves have increased, especially working with great numbers towards 1000. I think they really surprised themselves when they know they can do it (805, 2011).

In conclusion, success in higher levels of mathematics for young students is reliant on the initial mathematical experiences teachers provide in their classrooms. The results of our research suggest that there are four key factors that support the success of these young students:

1. Mathematics activities are hands-on with an opportunity for these young students to kinesthetically engage in the learning. These types of activities not only support the learning of mathematical concepts and development of language, but also create a sense of fun and enjoyment.

2. It is fundamental that initial learning activities are fail proof so that all students can have early success when participating in mathematics. This leads to confident, engaged mathematics learners.
3. Mathematics activities are differentiated so that students can extend their understanding of mathematics within an activity that they have already had success in. Rather than changing the context of the learning activity, students need to extend their learning by building on this initial success, and then transfer this learning to new contexts.
4. Conceptual learning is layered, and the key constructs that help students to link these layers are the representations and language used to explore the concepts. As we build conceptual understanding we also need to build the accompanying representations and language used. The ultimate goal is to build to abstraction. As students experience difficulties, stripping back the layers (conceptual, representational and language) helps them re-engage.

The focus of this chapter has been on identifying the particular dimensions that assisted these teachers and students to succeed in teaching and learning mathematics. In order to gain insights into the generalizability of these results to other countries and other contexts, academics from New Zealand and Canada were invited to share their perspectives on these dimensions. In all three countries the status of Indigenous people is significantly poorer than that of their non-Indigenous counterparts. Additionally their urban societies have been created by the British (and French), and built on the invasion of an Indigenous population.

New Zealand perspective by Roberta Hunter & Jodie Hunter[1]

In New Zealand Pāsifika and Māori form a disproportionally large group of the students who significantly underachieve in mathematics. This group of students also comprises more than 90 % of the students who attend schools termed low decile.[2] However, unlike the Australian context the majority of these high poverty schools are situated in the greater Auckland urban area. But in most other ways the demographics match the trend for teachers in marginalized contexts in Australia in that there are a greater percentage of recently qualified, less experienced teachers and a higher than normal proportion of overseas trained teachers who are under qualified to teach in these marginalized contexts.

Similar to the RoleM study twenty-three of the decile 1 schools in the greater Auckland urban area are involved in a New Zealand Ministry funded project titled *Developing Mathematical Inquiry Communities* (DMIC) and it is the teachers, students, and outcomes within this project we will be describing. DMIC takes a two-pronged approach involving professional development days and in class mentoring sessions. The project is focused on developing the teachers' pedagogical

---

[1]This contribution has been written by Associate Professor Roberta Hunter and Dr. Jodie Hunter from Massey University.

[2]Schools in New Zealand are ranked into deciles (low to high) as an indicator of the socio-economic level of the school community. The lowest decile ranking is a decile 1; the highest is decile 10. Students of Pasifika ethnicity predominantly attend schools within decile ratings of 1–3.

knowledge and pedagogical content knowledge, while also attending to their mathematical knowledge. Developing strong identities (through honoring the culture and language of the students) and positive mathematical dispositions is at the heart of the project. However, at the start of the project many of the participating teachers were overwhelmed by the contextual issues they faced. We frequently heard them voice statements like:

- these children come to school with no maths
- they do not know how to talk and they will not participate in talking
- they cannot ask questions because they do not know how.

We also viewed many examples of teachers wanting to always keep the students safe by reducing their expectations on them to communicate and limiting tasks to those they could complete with ease. We knew that to change these perceptions the teachers needed to see their students actively engaged in mathematical discourse and using a range of mathematical practices while they used high level mathematical reasoning to solve complex and challenging problems.

To achieve a change in teacher expectations and perceptions our immediate focus was placed on developing culturally responsive pedagogical ways to engage the students in mathematics. We embedded the social norms within core Pasifika values (for example, reciprocity, collectivism, family, respect, service) and used these to shape how students engaged in communicating their reasoning as they solved challenging and complex problems contextualized within the students' real world situations. As the RoleM study illustrated, when the students were able to access the problems their confidence grew, as did their teachers'.

Canadian perspective by Jacqueline Ottmann[3]

Interestingly, there may be more similarities than differences in the Aboriginal student experience in Australia and Canada, experiences that affect learning, in this case the learning of mathematics. The Canadian demographics and school completion statistics for Aboriginal peoples, which includes First Nations, Métis, and Inuit, has been drawing the attention of political and educational leadership, policy makers, curriculum developers, educators and researchers. First Nations, Métis and Inuit (particularly First Nations) peoples continue to be the fastest growing demographic. According to Statistics Canada (2011) Census, 1,400,685 people self-identified as Aboriginal—this is 4.3 % of the total Canadian population. The median age of Canada's Aboriginal population was 28 compared to 41 years of age for the general population. Forty-six percent of the total Aboriginal population was under the age of 25 and 28 % under the age of 14 years (Statistics Canada 2011). It is estimated that more than 600,000 Aboriginal youth will turn 15 between 2001 and 2026 (Hull 2005). In addition, Aboriginal peoples experience poverty at a greater level. 'The most poverty-stricken group of children are status First Nations children where 50 % live below the poverty line' (Macdonald and Wilson 2013,

[3]This contribution has been written by Associate Professor Jacqueline Ottmann from University of Calgary Werklund School of Education.

p. 12). The implications of poverty, residential school effects, unilateral and unsupportive legislations negatively influence learning. Macdonald and Wilson (2013) indicate that '60 % of First Nations parents have not completed high school' (p. 20). Teaching (knowledge, skills and attributes) is negatively impacted when the ecological systems, the lived experience of students is not investigated and understood.

The educational gap, this including mathematics, which exists in Australia also exists in Canada for Aboriginal students. The Auditor General in Canada (2011) stated that the education gap still exists between the general student population and First Nations students and that in some cases it has worsened. In the report *Decolonizing Our Schools: Aboriginal Education in the Toronto District School Board*, Dion et al. (2010) identified Aboriginal students as "invisible kids; the marginalized of the marginalized" (Brown 2012 n.p.). Dion et al. stated,

> The Toronto District School Board is failing to provide aboriginal students with the educational environment and experiences they require for success …"The board has not yet recognized that staff lack understanding (about aboriginal culture and history); the depth of ignorance plays a significant role in perpetuating the achievement gap" (Brown 2012). In addition to low school completion rates, Dion et al. noted, "Only 17 per cent of aboriginal students met the provincial standard in math in Grade 9 in 2007, compared to an average of 47 per cent of board students overall" (p. 16).

Neel (2007) supported the finding that overall, throughout Canada, K-12 student achievement in mathematics is much lower for Aboriginal students. Dr. Dion also explained that focused and intentional culturally-sensitive programming, curricula, lessons and teacher support, which was delivered by a pilot project, did make a difference to student engagement and academic success. The Banff International Research Station for Mathematical Innovation and Discovery (2015) research team asked:

> "We wonder, given the research around the different ways in which mathematics can be integrated into Aboriginal youth mathematical experiences and into school programs, why are these performance statistics not showing improvement?" (n.p.). Herein lies the challenge for educators.

As in Australia, it is difficult to attract and retain teachers in rural and isolated contexts, and obtaining quality resources can be an ongoing challenge (Wallin 2009). Many new teachers begin in these contexts then move on to larger centers, which typically has more resources. This kind of movement can send a powerful message of a lack of commitment to education and to students' learning to the whole school community. Teachers leave for many reasons, one may be what Leroy Little Bear calls "jagged worldviews colliding" (2000). Particularly in contexts that differ from a teacher's worldview, where teachers are unfamiliar with Aboriginal perspectives, it is important to reflect on the affective domain, one's values and belief systems. Ottmann and Pritchard (2010) explain:

> [C]lassroom practice is contextual, so there is no correct prescriptive method to teaching Aboriginal Perspectives. Meaningful learning happens when a teacher intimately knows himself or herself, and knows his or her students – where they come from and from what

> worldview they interpret their environment. Reaching this state of intimate understanding, may require deep, second-order change processes. In this respect, second-order change constitutes the seeking of personal understanding in relation to the environment. This process of discovery is a courageous movement, a shift to a place below the surface, to a more protective place, a place where the affective domain resides – a place where values and beliefs evolve. It is in a sense, a vision quest. If conscious, deliberate changes happen for an educator at this level, the chances of sustainability for those changes increase – the desired changes are captured for the future. (p. 41)

As in the RoleM study it is important to connect to both affective and cognitive learning domains to promote both teacher and consequently student development. When a teacher engages in exercises that touch both affective and cognitive aspects in an intentional on-going, process-oriented manner then dramatic first- and second-order changes in personal and professional development can occur, positively affecting every area of learning, again, this including the learning of mathematics.

References

Abedi, J., & Lord, C. (2001). The language factor in mathematics tests. *Applied Measurement in Education, 14*(3), 219–234.

Allard, A., & Santoro, N. (2004). Making sense of difference? Teaching identities in postmodern contexts. In P. Jeffery (Ed.), *Doing The Public Good: Positioning Educational Research (AARE 2004 International Education Research Conference Proceedings)* (pp. 1–20). Coldstream, VIC: Australian Association for Research in Education.

Australian Curriculum, Assessment and Reporting Authority. (2014). *NAPLAN achievement in reading, persuasive writing, language conventions and numeracy: National report for 2014.* Sydney, NSW: Author.

Australian Curriculum, Assessment and Reporting Authority (ACARA). (2009). *Shape of the Australian curriculum.* Barton, ACT: Commonwealth of Australia.

Bandura, A. (1991). Social cognitive theory of self-regulation. *Organizational Behavior and Human Decision Processes, 50*(2), 248–287.

Bandura, A. (1992). Self-efficacy mechanism in psychobiologic functioning. In R. Schwarzer (Ed.), *Self-efficacy: Thought control of action* (pp. 355–394). Washington, DC: Hemisphere.

Banff International Research Station for Mathematical Innovation and Discovery. (2015). Understanding relationship between Aboriginal knowledge systems, wisdom tradition, and mathematics: Research possibilities. Banff. Retrieved from http://www.birs.ca/events/2013/5-day-workshops/13w5120

Baumert, J., Kunter, M., Blum, W., Brunner, M., Voss, T., Jordan, A., et al. (2010). Teachers' mathematical knowledge, cognitive activation in the classroom, and student progress. *American Educational Research Journal, 47*(1), 133–180.

Boaler, J., & Staples, M. (2008). Creating mathematical futures through an equitable teaching approach: The case of railside school. *Teachers' College Record, 110*(3), 608–645.

Borman, G. D., & Kimball, S. M. (2013). Teacher quality and educational equality: Do teachers with higher standards- ratings close student achievement gaps? *The Elementary School Journal, 106*(1), 3–20.

Brown, L. (2012). Closing the 'achievement gap' for Toronto's Aboriginal students. Toronto: Toronto Star Newspapers Ltd. Retrieved from http://www.thestar.com/news/gta/2012/06/16/closing_the_achievement_gap_for_torontos_aboriginal_students.html

Cairney, T. (2003). Literacy within family life. In N. Hall, J. Larson, & J. Marsh (Eds.), *Handbook of early childhood literacy* (pp. 85–98). London: Sage Publications.

Celedón-Pattichis, S., & Turner, E. (2012). Explicame tu Respuesta: Supporting the development of mathematical discourse in emergent bilingual kindergarten students. *Bilingual Research Journal, 35*, 197–216.

Cooper, T. J., & Warren, E. (2008). The effect of different representations on Years 3 to 5 students' ability to generalise. *ZDM, 40*(1), 23–37.

Darling-Hammond, L., & Richardson, N. (2009). Teacher learning: What matters? *Educational Leadership, 66*(5), 46–53.

Denton, K., & West, J. (2002). *Children's reading and mathematics achievement in kindergarten and first grade*. Washington: National Center for Education Statistics, US Department of Education.

Dion, S., Johnston, K., & Rice, C. (2010). Decolonizing our schools: Aboriginal education in the Toronto District School Board. Toronto. Retrieved from http://ycec.edu.yorku.ca/files/2012/11/Decolonizing-Our-Schools.pdf. doi:10.3102/0002831204200237l

Government of Canada. (2011). 2011 June Status Report of the Auditor General of Canada. Retrieved from http://www.oag-bvg.gc.ca/internet/English/parl_oag_201106_04_e_35372.html

Hattie, J. A. (2009). *Visible learning: A synthesis of 800+ meta-analyses on achievement*. Abingdon: Routledge.

Hewitson, R. (2007). Climbing the educational mountain: A metaphor for real culture change for Indigenous students in remote schools. *Australian Journal of Indigenous Education, 36*, 6–20.

Hill, H. C., Rowan, B., & Ball, D. L. (2005). Effects of teachers' mathematical knowledge for teaching on student achievement. *American Educational Research Journal, 42*(2), 371–406.

Howard, P. (1997, December). *Aboriginal educators: Voices in our schools*. Paper presented at Annual Conference of the Australian Association for Research in Education. Brisbane, Qld: AARE. Retrieved from http://www.aare.edu.au/publications-database.php/2637/aboriginal-educators-voices-in-our-schools

Hull, J. (2005, June). Post-secondary education and labour market outcomes Canada, 2001. Ottawa, ON: Minister of Indian Affairs and Northern Development. Retrieved from http://publications.gc.ca/collections/Collection/R2-399-2001E.pdf

Jones, K., Kershaw, L., & Sparrow, L. (1995). *Aboriginal children learning mathematics*. Perth, Western Australia: MASTEC, Edith Cowan University.

Jorgensen, R., Grootenboer, P., Niesche, R., & Lerman, S. (2010). Challenges for teacher education: The mismatch between beliefs and practice in remote Indigenous contexts. *Asia-Pacific Journal of Teacher Education, 38*(2), 161–175. doi:10.1080/13598661003677580 .

Kent, A. M. (2004). Improving teacher quality through professional development. *Education, 124* (3), 427–435.

Kleickmann, T., Richter, D., Kunter, M., Elsner, J., Besser, M., Krauss, S., & Baumert, J. (2013). Teachers' content knowledge and pedagogical content knowledge the role of structural differences in teacher education. *Journal of Teacher Education, 64*(1), 90–106.

Krauss, S., Brunner, M., Kunter, M., Baumert, J., Blum, W., Neubrand, M., & Jordan, A. (2008). Pedagogical content knowledge and content knowledge of secondary mathematics teachers. *Journal of Educational Psychology, 100*(3), 716.

Linnenbrink, E. A., & Pintrich, P. R. (2003). The role of self-efficacy beliefs in student engagement and learning in the classroom. *Reading & Writing Quarterly, 19*(2), 119–137.

Little Bear, L. (2000). Jagged worldview colliding. In M. Battiste (Ed.), *Reclaiming Indigenous voice and vision* (pp. 78–85). Vancouver: UBC Press.

MacDonald, D. & Wilson, D. (2013). Poverty or prosperity: Indigenous children in Canada. Ottawa, ON: Canadian Centre for Policy Alternatives. Retrieved from https://www.policyalternatives.ca/publications/reports/poverty-or-prosperity

McDonald, S., Warren, E., & DeVries, E. (2011). Refocusing on oral language and rich representations to develop the early mathematical understandings of Indigenous students. *Australian Journal of Indigenous Education, 40*, 9–11.

Meaney, T., Fairhill, U., & Trinick, T. (2008). The role of language in ethnomathematics. *The Journal of Mathematics and Culture, 3*(1), 52–65.

Miller, J., & Warren, E. (2014). Exploring ESL students' understanding of mathematics in the early years: Factors that make a difference. *Mathematics Education Research Journal, 26*(4), 791–810.

Moschkovich, J. (2007). Using two languages when learning mathematics. *Educational Studies in Mathematics, 64*(2), 121–144.

Moyer, P. (2000). Communicating mathematically: Children's literature as a natural connection. *The Reading Teacher, 54*(3), 246–255.

Nasir, N. I. (2011). *Racialized identities: Race and achievement among African American youth.* Stanford University Press.

National Council of Teachers of Mathematics (NCTM). (2000). *Principles and standards for school mathematics*. Reston, VA: Author.

Neel, K. I. S. (2007). Numeracy in Haida Gwaii, BC: Connecting community, pedagogy, and epistemology (pp. 313). Unpublished doctoral dissertation, Simon Fraser University.

Ottmann, J., & Pritchard, L. (2010). Aboriginal perspectives and the social studies curriculum. First Nations Perspectives: The Journal of the Manitoba First Nations Education Resource Centre Inc., *3*, 21–46. Retrieved from http://www.mfnerc.org/wp-content/uploads/2012/11/5_OttmanPritchard.pdf

Shaftel, J., Belton-Kocher, E., Glasnapp, D., & Poggio, J. (2006). The impact of language characteristics in mathematics test items on the performance of english language learners and students with disabilities. *Educational Assessment, 11*(2), 105–126.

Spanos, G., Rhodes, N. C., Dale, T. C., & Crandall, J. (1988). Linguistic features of mathematical problem solving: Insights and applications. In R. R. Cocking & J. P. Mestre (Eds.), *Linguistic and cultural differences on learning mathematics* (pp. 221–240). Hillsdale: Erlbaum.

Statistics Canada. (2011). Aboriginal peoples in Canada: First Nations, Métis and Inuit. Retrieved from http://www12.statcan.gc.ca/nhs-enm/2011/as-sa/99-011-x/99-011-x2011001-eng.cfm

Thorpe, K. J., Tayler, C. P., Bridgstock, R. S., Grieshaber, S. J., Skoien, P. V., Danby, S. J., & Petriwskyj, A. (2004). *Preparing for school: Report of the Queensland preparing for school trials 2003/4*. Brisbane, QLD: School of Education QUT.

Valle, M. S., Waxman, H. C., Diaz, Z., & Padrón, Y. N. (2013). Classroom instruction and the mathematics achievement of non-english learners and english learners. *The Journal of Educational Research, 106*(3), 173–182. doi:10.1080/00220671.2012.687789.

Wallin, D. (2009). Rural education: A review of provincial and territorial initiatives. Ottawa: Canadian Council on Learning. Retrieved from http://www.ccl-cca.ca/pdfs/OtherReports/RuralEducation.pdf

Warren, E., & Cooper, T. J. (2009). Developing mathematics understanding and abstraction: The case of equivalence in the elementary years. *Mathematics Education Research Journal, 21*(2), 76–95.

White, S., & Reid, J. (2008). Placing teachers? Sustaining rural schooling through place consciousness in teacher education. *Journal of Research in Rural Education, 23*(7), 1–11.

Chapter 6
Redressing the Imbalance

Abstract This chapter reviews the main findings of the study in relation to the research questions, and presents the implications for practice and research. While participating in the RPL, we theorize that teachers and students transitioned through five distinct stages of a professional learning and teaching trajectory. Each stage contributed to their journey towards becoming expert teachers and their students achieving academically, and each stage required different levels of support. Given the size and scope of the participants and the differing contexts in which they were situated, we also claim that this trajectory is generalizable to many other marginalized contexts at both the national and international level. Thus, the hypothesized professional learning and teaching trajectory frames our recommendations with regard to assisting young ESL students and Indigenous students living in disadvantaged contexts to effectively engage in Western mathematics.

Professional Learning and Teaching Trajectory for Marginalized Contexts

A meta-analysis of the data resulted in the development of a professional learning and teaching trajectory for effective learning and teaching in marginalized contexts. Our original model focused purely on the teachers and the changes they made during the professional learning (Warren and Miller 2013). Underpinning the development of this new trajectory is the stance that teaching and learning are closely related constructs. A change in one leads to a change in the other. As teachers transition through the stages of the professional learning and teaching trajectory, they broaden and deepen their understandings of mathematics, and change the way in which they deliver mathematics lessons. Subsequently they recognize that these changes have a positive impact on their students' learning. As a consequence of these changes, students in our project had (a) increased their mathematical content knowledge and mathematical language (b) made connections

E. Warren and J. Miller, *Mathematics at the Margins*,
SpringerBriefs in Education, DOI 10.1007/978-981-10-0703-3_6

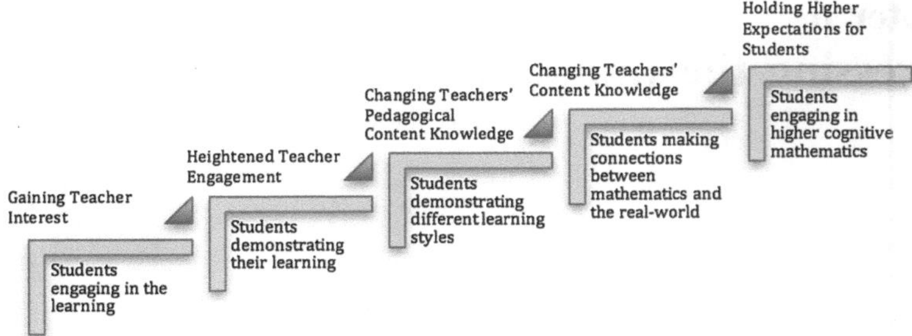

Fig. 6.1 RoleM professional learning and teaching trajectory

between their mathematical knowledge and real-world contexts, and (c) engaged in higher order thinking. Each stage influences the next, for example teachers' effective teaching influences students' effective learning, which then in turn influences the teachers' teaching.

Figure 6.1 presents the theorized professional learning and teaching trajectory. This trajectory builds on the trajectories presented in Chaps. 3 and 4 that emerged from the teacher and student data. It includes the stages that both teachers and students progress through as they transition to become expert teachers and highly engaged learners of mathematics in marginalized contexts. In addition it begins to answer the overarching question:

What Is the Interplay Between Effective Teaching and Effective Learning of Mathematics in These Contexts?

The initial stage focuses on *gaining teachers' interests.* It was necessary for teachers to have an immediate increased interest in teaching mathematics. Providing teachers with quality hands-on resources and mathematical learning experiences for their year levels helped 'gain their attention'. Often there are limited resources in marginalized schools and by providing teachers with quality hands-on resources this barrier was removed. Additionally, the professional learning included demonstrations by experts of how to use the hands-on resources and implement the learning experiences in their classrooms. Further support was given beyond the initial PL day as each learning experience was accompanied with a video that teachers could download and watch. The video contained examples of experts demonstrating teaching these lessons to small cohorts of students. This provided an opportunity for teachers to revisit ideas presented at the PL days. These aspects are particularly crucial for gaining the interest of teachers working in marginalized contexts, as there is often high staff turnover and minimal expertise for teachers to call upon (Lyons et al. 2006).

This initial stage for students focuses on *students engaging in the learning*. Prior to RoleM, mathematics lessons often consisted of teachers standing at the front of the class endeavoring to deliver the content, with the expectation that their students would copy the work provided. This often resulted in a climate where students were simply not listening nor were they interested. As teachers started to use the RoleM learning experiences, students had opportunities to engage with purposely chosen hands-on resources that were deemed to be culturally appropriate and focused on core mathematical concepts. Students worked together in small groups engaging with the materials. The resources were underpinned by mathematical structures but were delivered as fun learning experiences. As a result, students became more task-focused as they 'played' with the mathematics. Student engagement is conjectured to consist of three components: learning-related work habits (persistence at tasks, completing tasking); cognitive behaviors (problem solving); and emotions (enthusiasm, interest) (Newmann 1992). We suggest that at this first level, students' engagement was predominantly emotional.

The second stage involves *heightened teachers' engagement*. As a result of students' emotional engagement teachers became more confident to independently trial the RoleM activities and resources in their classroom environments. As they trialed the activities they observed their students increasing engagement and focus on mathematical learning. As a result of this, teachers used the RoleM materials more regularly during their mathematics lessons. Consequently, teachers were beginning to buy into changing the ways they teach mathematics to these students, beginning to change their pedagogical practices, and in particular their knowledge of content and teaching (Ball et al. 2008).

This second stage for students involves *students demonstrating their learning*. Due to the RoleM learning experiences being used more readily in classrooms, students were emotionally engaging with mathematics on a more regular basis in class. This provided students with greater opportunities to exhibit their learning as they became more familiar and confident with the mathematics, that is, demonstrate their cognitive engagement (Newmann 1992). Additionally, students were exploring different representations of mathematics concepts as teachers trialed ideas. Consequently, students began to reveal a diverse range of needs.

The third stage is *changing teachers' pedagogical content knowledge*. During this stage experts worked in the individual teacher's classrooms to further support and assist teachers to implement RoleM. Experts modelling the mathematics lessons had a substantial impact on teachers. This modelling provided teachers with the opportunity to draw on the experts' experiences and re-engage with the resources/activities that they were experiencing difficulty with in their own classroom context. With ongoing support, teachers and experts became co-constructors of knowledge moving within and beyond each other's Zone of Proximal Development (Vygotsky 1978). This stage further impacted on the delivery of mathematics. There was a marked shift away from teachers' use of worksheets as a means to teach mathematics. Teachers had begun to differentiate the learning experiences to cater for individual student needs. This challenged teachers' practices and provided insights into what these students were capable of achieving, going against the belief that these students are not capable of achieving in mathematics (Hewiston 2007).

This third stage for students is *students demonstrating different learning styles.* As a consequence of teachers providing a range of different pedagogical approaches to their students, they began to observe that *all* their students were learning. We conjecture that this stage is where the two constructs (teaching and learning) are in symphony. While the relationship between cognitively challenging instructional tasks and high levels of student thinking has been acknowledged in the literature (Hiebert and Wearne 1993; Stein and Lane 1996), the findings of this research suggest the teaching and learning relationship is also evident when we use a diversity of pedagogical practices. The more diverse the teaching practices, the greater the styles of learning students exhibit. The notion of symphony suggests that one does not result in the other but that they work in harmony. As we notice different ways students learn, we further diversify our teaching practices. As we diversify our teaching practices, we see more styles of learning.

The fourth stage is *changing teachers' content knowledge.* As teachers engaged in the RPL, they increased their own mathematical content knowledge. As teachers increased their own knowledge of mathematics, they gained new 'tools' for teaching. Teachers were able to provide high achieving students with more challenging tasks. Additionally, teachers were able to identify the common mathematics misconceptions that students demonstrated and appropriately cater for these students in their everyday teaching. The ease with which teachers could adapt leaning activities during the lessons increased over time. As a result, teachers could provide on the spot adaptations to individually cater for all students. These traits are beginning signs of an expert teacher: those teachers whose knowledge is integrated, and who are flexible in its use in the classroom (Borko and Livingston 1989). Expert teachers know what to teach, and how to structure and organise this for students' context (Askew 2008). Peer sharing was also evident as teachers began to discuss their successes and failures with other colleagues, demonstrating reflective practice.

The fourth stage for students is *making connections between mathematics and the real world.* As students become more competent with their mathematical knowledge and understanding, they become more capable of transferring this knowledge. Transfer of knowledge occurs when the learning that has taken place in the first learning situation influences the capability of performing in the second learning situation (Thorndike 1906/1913). Students were able to transfer the mathematical knowledge they had learnt in a particular RoleM learning experience to assist them answer real-world mathematics problems posed by their teachers. Past studies have indicated that real-life mathematics tasks are more difficult for students who are from working class contexts (Cooper and Dunne 2000; Lubienski 2000). It is conjectured that because the learning experiences were rich in representations that were familiar to students' context, they provided a platform for these students to transfer their knowledge and make connections to the real world. Additionally, students also were making connections or transferring their knowledge beyond the classroom environment. There were many instances where students used their mathematical language in play situations outside the classroom. They were able to use their acquired mathematics skills to assist them in everyday life. This is an important step for students' numeracy development (COAG 2008; NCTM 2000; OECD 2009).

The final stage is teachers *holding higher expectations for their students*. At this stage, teachers were able to identify the effects that their teaching had with respect to students' mathematical learning. Teachers became less focused on the factors that are often attributed to students not achieving in these marginalized contexts (e.g., external school factors, absenteeism, behavior, language), and were no longer using these excuses for students not making gains. It was evident that teachers (a) were taking ownership of their lessons and students learning, and (b) were more flexible as the needs or goals of students emerged from the mathematics lessons. Both of these are qualities that are also essential to expert teachers (Borko and Livingston 1989). Setting high expectations for students from disadvantaged contexts is the pinnacle of this project. It resulted in teachers providing effective teaching practices that resulted in students achieving in mathematics at levels equivalent to their peers.

The final stage for students is *students engaging with higher cognitive levels of mathematics*. As students moved through the RoleM learning experiences, they acquired a repertoire of mathematical skills. Their mathematical conceptual understanding and mathematical language use increased. In addition, their understanding of underlying mathematical structures became more evident. These understandings contributed to students engaging with and exhibiting higher levels of mathematics. The flexibility in the learning experiences meant that as students were successful in a particular learning experience, teachers were able to easily challenge students by extending the activity (e.g., changing the place value from whole numbers to decimals, or extending to 1000s).

How Do Teachers' Dispositions Towards Teaching Mathematics Influence Their Classroom Practice in These Contexts?

Teachers' dispositions to teaching mathematics have a profound influence on their classroom practice in these contexts. As evidenced by the theorized professional learning and teaching trajectory, as their confidence in teaching mathematics became more positive, their use of different pedagogies in the classroom increased. We are suggesting that a change in teachers' dispositions towards teaching mathematics is a two-fold process. First, teachers need to see that as they, themselves, change, their students become more positively engaged with the mathematics they are teaching. Second, it is essential for them to realize that these changes result in greater student learning.

How Does Teachers' Mathematics Content Knowledge and Pedagogical Knowledge Influence Their Classroom Practices in These Contexts?

Once teachers gain confidence in teaching mathematics, they begin to identify the need to change how they teach mathematics and to increase their own knowledge of mathematics. The results of this research suggests that in these contexts this initial focus is on increasing their pedagogical knowledge, ascertaining a variety of ways they can teach mathematics. This results in teachers being able to cater for the diverse range of learners in their classroom. It also results in more students being able to exhibit their learning. As these students become more cognitively engaged, then the need to extend and challenge their learning emerges. It is at this point that teachers' MCK is required. If it is lacking, then it impacts on their ability to (a) provide activities that are high level activities, (b) differentiate learning experiences, (c) understand what representations support students' learning of particular mathematics concepts, and (d) provide opportunities for linking the mathematical language to the representation. Our findings suggest that teachers recognize when this point is reached and willingly ask for help. The difficulty is that often there is no expert in the community for them to call on. It is at this point that professional learning support is crucial.

How Does Professional Learning Best Support the Development of This Interplay?

Based on the data and our own reflections, we argue that professional learning requires two main focuses: the enactment of mathematics and the acquisition of mathematics. Both of these focuses are represented in Fig. 6.1. The enactment is the teaching side of the trajectory and the acquisition is the student learning side of the trajectory. As evidenced by the discussion that accompanied this professional learning and teaching trajectory, we also conjecture that these are interconnected and, in some instances, symbiotic focuses. Each step in the trajectory represents a movement backwards and forwards between the two elements illustrated in Chap. 1, Fig. 1.1, with the affective domain and cognitive domain embedded within the steps. In addition, these recommendations expand previous implications and recommendations drawn from our research (Warren and Miller 2013; Warren and Quine 2013). Figure 6.2 begins to tease out these relationships.

The results from our research suggest that the predominant focus of change is initially on the affective domain, with the cognitive domain playing a secondary role. As teachers and students move through the change, the affective domain begins to fade into the background and the cognitive domain comes into the fore. Thus, the accompanying professional learning opportunities need to be in tune with these movements and progress accordingly. The recommendations we are making for

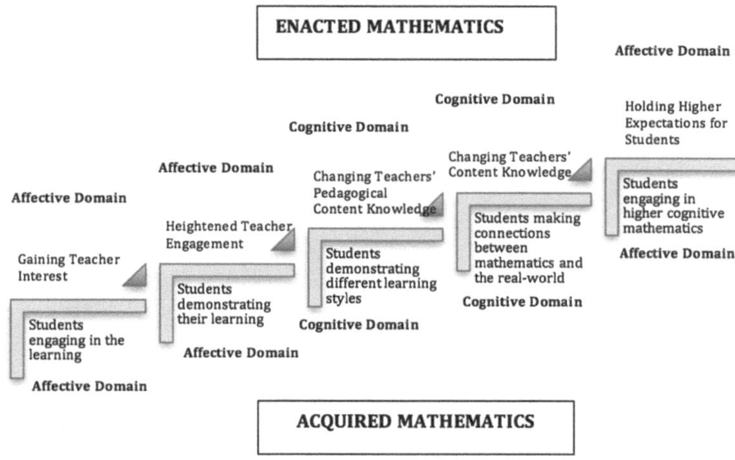

Fig. 6.2 The interplay of the development of teaching and learning

enhancing the enacted mathematics and attainment of mathematics in marginalized contexts are presented as three interconnected phases of change: beginning, continuing and deepening.

Implications and Recommendations for Enacting and Acquiring Mathematics

Beginning the Journey

Building teachers' and students' confidence

Developing knowledge of the language of mathematics and its use orally.

The recurring issue for teachers working in these communities is the gap between the language of the students, the language of instruction, and the language of mathematics. Emphasizing teachers' use of oral language in conjunction with rich mathematical representations ensures teachers and students develop a shared register of communication.

Developing an understanding of how to effectively use proven mathematical learning experiences.

Providing teachers with contextually appropriate learning experiences that can be used immediately, and students can experience instantaneous success with, removes the feeling that they are unprepared and under-resourced. Trialing proven experiences also assists teachers to feel more confident about their ability to teach mathematics in these contexts.

Supporting teachers in their classrooms with on-site visits by experts.

These visits need to be responsive to teachers' specific requests and focus on helping students to exhibit their understanding of mathematics. Teachers' and students' confidence builds when students are engaged in learning mathematics and are experiencing success.

Continuing the Journey

Increasing teachers' and students' knowledge of mathematics.

Gaining more general mathematics pedagogical knowledge.

Catering for the diverse range of students in marginalized contexts entails learning how to confidently use differing ways of teaching mathematics. This involves using an array of representations, language patterns, and targeting different styles of learning. This not only helps teachers gain a deeper understanding of how students learn but also allows students to exhibit their potential to learn.

Gaining a deeper understanding of how to differentiate the learning.

Being able to differentiate the learning is not necessarily about providing different types of activities for different levels of learners. It is more about understanding how concepts progressively develop, and using this knowledge to gear up or strip back a particular learning experience as required. This process helps to address the wide range of student abilities that exists in many disadvantaged communities and allows all students to experience success.

Deepening the Journey

Increasing the expectations for teachers and students.

Gaining a deeper understanding of mathematical content knowledge.

In order to set higher expectations for their students, teachers need to have a deeper knowledge of mathematical concepts, conceptual frameworks, and learning trajectories. This assists teachers to understand the hierarchical nature of mathematics, how understanding builds on prior understanding, and provides ways to use this knowledge to target students' learning.

Gaining a deeper understanding of mathematical pedagogical knowledge.

The ability to anticipate, plan and improvise is at the heart of teaching mathematics effectively. Having the ability to be flexible in how mathematics can be taught includes understanding the different ways that mathematical concepts can be represented and how these representations are connected. It also includes knowledge about what makes a subject cognitively easy or difficult for student learning, as well as knowledge of misconceptions and pre-conceptions students may hold about mathematical concepts.

For Indigenous Contexts

Gaining a deeper understanding of the context by working with Indigenous education officers.

Building strong relationships with Indigenous teacher assistants and Indigenous leaders in the community is essential for teachers to begin to develop a deeper understanding of the context they are working in.

Implications for the Professional Learning of Teachers Working in These Contexts

As evidenced by the results of RoleM, professional learning opportunities that are ongoing, contextualized and emphasizing teachers' pedagogical and content knowledge are more likely to enhance the learning of both teachers and students. In addition providing effective culturally appropriate resources clearly linked to positive student outcomes further enhances this learning. But it also needs to be acknowledged that this learning is a slow process. It was in the second year of participation that student learning became the primary focus of participating teachers. It was only after this period that teachers saw themselves as capable of effectively structuring and organizing the learning to cater for the diversity of students in their classroom, the hallmarks of effective teachers (Hattie 2003).

We also hypothesize that in these contexts where teachers are constantly moving in and out, an approach of focusing on a new subject area each year is ineffective. Such an approach inhibits teachers' and students' progression along the professional learning and teaching trajectory. They are continually returning to the beginning phases. We conjecture that a more effective approach is an ongoing focus on the key subject areas of literacy and mathematics, the two areas that are known to underpin future employment and further education opportunities (Lamb and McKenzie 2001).

New Zealand Perspective by Roberta Hunter and Jodie Hunter[1]
In a similar way to the RoleM study, the teachers in the Developing Mathematical Inquiry Communities (DMIC) project transitioned through a professional teaching and learning trajectory. However, unlike the RoleM study, the participating teachers were not provided with resources or activities although contextualized problems situated in the real world of the students were developed and explored during workshops with the teachers. Some workshops also directly examined and explored specific aspects of mathematical knowledge with the intention of strengthening teacher understandings. Particular focus was placed on developing teacher

---

[1]This contribution has been written by Associate Professor Roberta Hunter and Dr Jodie Hunter from Massey University.

knowledge of Fractions, Decimals, and Percentages; Ratios; Geometry and Early algebraic reasoning. An additional tool the teachers were provided with was a Communication and Participation Framework. This tool provided a number of proposed pedagogical actions to gradually induct the Pāsifika and Māori students into use of a range of mathematical practices within a mathematical inquiry community. Inherent in the tool was a trajectory, which shifted teacher focus from students providing mathematical explanations to drawing on representations and mathematical language to engage in mathematical justification, and argumentation while also attending to the social norms of the classroom in culturally responsive ways.

As Warren and Miller explain, teaching and learning are closely connected constructs. It was evident in our findings, as it is in theirs, that as one aspect of the teachers' practices changed so too did other aspects. In our case this was a growth in pedagogical strategies to engage the Pāsifika and Māori students in talking about how they were doing mathematics. As student voice increased so too did the teachers' skills at using dynamic assessment to respond in appropriate ways. This in turn positively impacted in many different ways on the students' learning. As our teachers saw what their students were really capable of achieving their expectations of higher levels of achievement and engagement in mathematical practices increased.

Our results concur with that illustrated in Fig. 6.1 with a trajectory, which followed a similar pathway. The immediate challenge was to gain teacher and student interest and through this to change perceptions of lowered expectations in both the teachers and students. Our tool focused on positioning this group of marginalized learners to access the mathematical discourse while engaged in tasks, which were differentiated with multiple entry and exit points. A goal was to reverse the balance of teacher student talk so that the teacher voice did not dominate as it had in former mathematics lessons where the Pāsifika and Māori students most often played the role of polite listeners. As student voice increased so too did the need for teachers to employ a more facilitative role and shift towards being a member of the learning community rather than the leader. At the same time, the teachers recognized a need for deeper and richer mathematical knowledge to connect student reasoning to the big mathematical ideas, which in turn led to higher order thinking in the students. Clearly evident is the cyclical shape, which begins and ends with changing teacher beliefs.

Canadian Perspective by Jacqueline Ottmann[2]

The RoleM professional development process provided teachers with opportunities to support their learning and development. They engaged in professional development that was process-oriented, and they highlighted the inextricable connection between teaching and learning. The Western and Northern Canadian Protocol

[2]This contribution has been written by Associate Professor Jacqueline Ottmann, University of Calgary Werklund School of Education.

(WNCP) (2013) (which included Aboriginal director representation Manitoba, Saskatchewan, Alberta, Yukon, Nunavut and the North West Territories) recognized the importance of this when the engaged in the Charter 3: "Our Way is a Valid Way": Professional Educator Resource project. The WNCP goal was to support teacher knowledge, skill and affective development in Aboriginal history and perspectives so student learning could be positively impacted. The group recognized the importance of supporting teacher learning in a holistic way.

Donald, Glanfield and Sterenberg (2011) contemplated "on what it means to do [mathematics] research in and with an Indigenous community" (p. 72). They came up with three stances to conducting mathematics research, and we would posit, stances to the teaching of mathematics education to Aboriginal students: "a mathematically deficit response, a culturally deficit response and a culturally relational response" (p. 72). The mathematically deficit response demands, "If you adopt program X that has proven to be successful in another location, then achievement in mathematics in your school will improve" (p. 74). This focuses on the knowledge deficit. The danger in a culturally deficit stance is that it is seen as the "restorative tool" (p. 76), the key to "changing educational experiences of Indigenous peoples" (p. 76), and it "distracts from and minimizes the historic and ongoing systemic racism and discrimination that have affected Aboriginal peoples" (p. 76). Culturally relational stance is "messy" (p. 80), more time consuming because it requires active listening, and observing and collaborative work. Donald et al. (2011) explain:

> We are with the community in multiple ways. We work alongside one another and with the community and school staff to question the ways in which the notion of culture is typically taken up… we challenge our assumptions about what the culture says the students should be and we are coming to understand and honour the philosophies that underlie the culture of the community. (p. 80)

The authors describe the culturally relational stance as emergent and generative, it is "where knowing arises in action because of the highly relational nature of learning and social interaction" (p. 80). Donald's et al. (2011) research supports the RoleM professional development process.

Increasingly, there are educational initiatives that are emerging to support Aboriginal student learning in mathematics. These include:

1. STEM (Science, technology, engineering, mathematics) programs specifically for Aboriginal students and provides support for teacher learning (e.g., National Aboriginal Outreach Program by Actua 2012).
2. The First Nation Student Success Program (FNSSP) (Saskatoon Tribal Council 2012), a proposal-driven program, "supports projects that increase students' achievement levels in reading and writing (literacy), mathematics (numeracy), and encourages students to remain in school (student retention)" (Aboriginal Affairs and Northern Development Canada 2014 n.p.).
3. The Math Catcher: Mathematics through Aboriginal Storytelling website (Jungic and MacLean 2011) is the result of a research project that explored mathematics and science learning of Aboriginal students. The resources within

the website provides teachers, "with the opportunity to explore ways to demystify mathematics for … students through storytelling and hands-on activities" (n.p.). The research findings included two conclusions "as strategies for overcoming challenges in teaching mathematics to Aboriginal youth: teach math in a cultural context of the students, and teach basic skills and problem-solving early" (n.p.).

4. The University of Calgary Galileo Educational Centre (2008) has done work that has focused on "reflexive pedagogy and visible learning", and has provided examples of what Donald et al. (2011) identify as a relational stance of learning alongside two local First Nations communities on two separate projects (e.g., http://www.galileo.org/initiatives/moka-meyo/index.html).

When teaching Aboriginal students, it is so important to consider what the community already knows, and has known for some time, about mathematics (Mann 2006; Aikenhead and Michell 2012). The knowledge and practice of mathematics from an Indigenous perspective can provide meaning and greatly strengthen the learning for both teacher and student. At any rate, learning mathematics should be fun and experiential for both teacher and student! Gitchi Meegwich!

References

Aboriginal Affairs and Northern Development Canada. (2014). First Nation Student Success Program. Retrieved from https://www.aadnc-aandc.gc.ca/eng/1100100033703/1100100033704.

Actua. (2012). National Aboriginal Outreach Program. Retrieved from http://www.actua.ca/aboriginal/.

Aikenhead, G., & Michell, H. (2012). *Bridging cultures: Indigenous and scientific ways of knowing nature*. Toronto: Pearson.

Askew, M. (2008). Mathematical discipline knowledge requirements for prospective primary teachers, and the structure and teaching approaches of programs designed to develop that knowledge. In K. Krainer & T. Wood (Eds.), *The international handbook of mathematics teacher education* (Vol. 1, pp. 13–36). Rotterdam: Sense Publishers.

Ball, D. L., Thames, M. H., & Phelps, G. (2008). Content knowledge for teaching what makes it special? *Journal of Teacher Education, 59*(5), 389–407.

Borko, H., & Livingston, C. (1989). Cognition and improvisation: Differences in mathematics instruction by expert and novice teachers. *American Educational Research Journal, 26*(4), 473–498. doi:10.3102/00028312026004473.

Cooper, B., & Dunne, M. (2000). *Assessing children's mathematical knowledge: Social class, sex and problem-solving*. Buckingham: Open University Press.

Council of Australian Governments (COAG). (2008). *National numeracy review report*. Canberra, ACT: Commonwealth of Australia.

Donald, D., Glanfield, F., & Sterenberg, G. (2011). Culturally relational education in and with an Indigenous community. *In education, 17*(3), 72–83.

Galileo Educational Network. (2008). Mokakioyis Meyopimatisiwin. Retrieved from http://www.galileo.org/initiatives/moka-meyo/index.html.

Hattie, J. A. (2003, October). Teachers make a difference: What is the research evidence? In *Background paper to invited address presented at the 2003 ACER Research Conference, Melbourne, Australia*. Retrieved from http://www.acer.edu.au/documents/TeachersMakea-DifferenceHattie.doc.

Hewitson, R. (2007). Climbing the educational mountain: a metaphor for real culture change for Indigenous students in remote schools. *Australian Journal of Indigenous Education, 36*, 6–20.

Hiebert, J., & Wearne, D. (1993). Instructional tasks, classroom discourse, and students' learning in second-grade arithmetic. *American Educational Research Journal, 30*(2), 393–425.

Jungic, V. & MacLean, M. (2011). The math catcher: mathematics through aboriginal storytelling. Burnaby, BC: Simon Fraser University. Retrieved from http://mathcatcher.irmacs.sfu.ca/.

Lamb, S., & McKenzie, P. (2001). *Patterns of success and failure in the transition from school to work in Australia, LSAY Report 18*. Melbourne, VIC: ACER.

Lubienski, S. T. (2000). Problem solving as a means toward mathematics for all: An exploratory look through a class lens. *Journal for Research in Mathematics Education, 31*(4), 454–482.

Lyons, T., Cooksey, R., Panizzon, D., Parnell, A., & Pegg, J. (2006). *Science, ICT and mathematics education in rural and regional Australia: The SiMERR national survey*. Canberra, ACT: Department of Education, Science and Training.

Mann, C. (2006). *1491: New revelations of the Americas before Columbus*. New York: Vintage Books.

National Council of Teachers of Mathematics (NCTM). (2000). *Principles and standards for school mathematics*. Reston, VA: Author.

Newmann, F. M. (1992). *Student engagement and achievement in American secondary schools*. New York: Teachers College Press.

Organisation for Economic Co-operation and Development (OECD). (2009). *Mathematical literacy*. Retrieved from http://www.oecd.org/dataoecd/38/51/33707192.pdf.

Saskatoon Tribal Council. (2012). First Nations student success program. Retrieved from http://www.sktc.sk.ca/fileadmin/user_upload/docs/FIRST_NATION_STUDENT_SUCCESS_PROGRAM_Website.pdf.

Stein, M. K., & Lane, S. (1996). Instructional tasks and the development of student capacity to think and reason: An analysis of the relationship between teaching and learning in a reform mathematics project. *Educational Research and Evaluation, 2*(1), 50–80.

Thorndike, E. L. (1913). The influence of improvement in one mental function upon the efficiency of other functions. In R. F. Grose & R. C. Birney (Eds.), *The transfer of learning: an enduring problem in psychology: selected readings* (pp. 1–6). Princeton, NJ: D. Van Nostrand.

Vygotsky, L. S. (1978). *Mind in society: The development of higher psychological processes*. Cambridge, MA: Harvard University Press.

Warren, E., & Miller, J. (2013). Enriching the professional learning of early years teachers in disadvantaged contexts: The impact of quality resources and quality professional learning. *Australian Journal of Teacher Education, 38*(7), 91–111.

Warren, E., & Quine, J. (2013). Enhancing teacher professional development for early years mathematics teachers working in disadvantaged contexts. In L. English & J. Mulligan (Eds.), *Reconceptualizing early mathematics learning: advanced series in mathematics education* (pp. 283–307). Netherlands: Springer.

Western Canadian Northern Protocol. (2013). "Our Way is a Valid Way": Professional Educator Resource. Retrieved from http://www.yesnet.yk.ca/firstnations/pdf/13-14/our_way_resource.pdf.

Chapter 7
Maintaining the Momentum

Abstract This chapter discusses the importance of maintaining the momentum of the RoleM project (and other like projects), and gaining its sustainability for students and teachers in marginalized contexts. New Zealand and Canadian perspectives are presented in relation to actions that ensure a smooth transition for these young children throughout their first four years of school. Differences and similarities are drawn between all contents to present key factors that support sustainability of programs in schools. Finally, conclusions are drawn, and actions delineated that ensure a smooth transition for these young students throughout their first four years of school.

Maintaining the momentum of an initiative entails teachers' commitment to continually engage in improving their teaching and their students' learning after professional development ceases. This commitment is essential to ensuring long-term successful outcomes for both staff and students. Thus teachers' focus needs to be on the long-lasting impact of an initiative, rather than short-term goals or gains. Often teachers implement new practices after professional development ceases, but these new practices are short lived (Timperley 2008). Teachers soon return to practices they used prior to the professional development occurring. We contest that the continuation and adaptation of programs, such as RoleM, needs to be a whole-school commitment, one that fosters quality teacher practices and continued improvement in students' outcomes.

Sustainability is the ability, of a school, to continue a program/initiative without ongoing support from external partners (Knight 2005). Sustainable programs in schools are about moving from the implementation phase, trialing new ideas or programs, to the institutionalization phase, when the program is embedded into teachers' practice (Anderson and Stiegelbauer 1994). Anderson and Stiegelbauer (1994, p. 280) highlight that institutionalized change is greatest under the following conditions:

E. Warren and J. Miller, *Mathematics at the Margins*,
SpringerBriefs in Education, DOI 10.1007/978-981-10-0703-3_7

- Solving a problem is significant to the initiated change;
- Local priorities are central;
- Administrators are committed to the change and provide ongoing support, assistance, incentives, and pressure;
- Increasing number of teachers and administrators engage with the planning, developing and refining;
- Effects can be demonstrated and monitored; and
- Creation of organizational structures to support the continuation of the program.

Additionally, maintaining the momentum is about creating capacity and opportunity for sustainability. This is dependent on both the model and delivery of the professional learning, and the organizational structures that are in place to assist teachers once support is withdrawn (Timperley 2008). Thus, the sustainability of a program needs be considered during the implementation of professional learning, rather than considered post program delivery (Timperley et al. 2007).

RoleM drew on past research (e.g., Earl et al. 2003; Hargreaves and Fink 2003; Timperley et al. 2007; Warren and Miller 2013; Warren and Quine 2013) to ensure that the model of professional learning established provided the potential for sustainability. Three key features underpinned this notion. The first entailed establishing and building teachers' skills in mathematics to equip them with best practices for teaching and learning beyond the professional learning days. Focusing on both teachers' MCK and PCK impacted the practice of teachers, and led to them developing new knowledge and practices post the professional learning. The knowledge that was delivered through the professional learning had depth (Earl et al. 2003), and was embedded in theories of mathematical teaching and learning. Additionally, the professional learning was more than just an 'add on' to the teachers' existing workload. It was integrated into the current curriculum making it easy to implement using current classroom practices.

The second involved investing in building long-term capacity for improvement within the school (Hargreaves and Fink 2003). This is about having long-term targets for both staff and students. In the early years of school the targets are on successful transitions from home to school, with a focus on students' engagement in school mathematics. The longitudinal approach to creating and implementing the RoleM activities ensured the successful transition of students through the first four years of school. As RoleM followed students from Foundation to Year 3 it was possible for staff and researchers to set, and measure long-term goals for students. Furthermore, teachers continued professional learning through on-site visits from experts established supported building long-term improvement both for themselves and their students.

Finally, support for sustaining the project needs to be established within the wider community and with school leaders. This creates a sense of commitment in the school community. The role of leadership is central to sustaining change in teachers' practice. Principals and mathematics curriculum leaders are key players in sustaining professional learning (Millett et al. 2004; Hargreaves and Fink 2003). Additionally, sustaining programs in schools creates a sense of stability for both staff and students, and sustainability of programs reaches beyond teachers accessing the professional learning. RoleM was committed to a whole-school approach. It not only offered professional learning opportunities to the participating teachers, but also offered professional learning to principals (not included in this book), curriculum leaders and the Indigenous teacher assistants within schools. This ensured that all four groups had a common basis from which to plan for the future sustainability of their students' learning of mathematics.

Barriers for Sustainability

Creating sustainable environments is challenging for schools in marginalized contexts. With continued teacher and leadership change it is difficult to maintain stability of a project past the initial stages (Fink 2000). As teachers gain an understanding of how to support the learning of students in these contexts they commonly either leave the school within their first year or leave at the conclusion of this first or second year. All the knowledge gained from the professional learning is in essence 'lost'. Additionally, coaching/mentoring support is not present in the school due to this high turnover of staff. Hence, to overcome these issues it is essential for initiatives, such as the RoleM, to build long-term relationships with schools. This helps to ensure their sustainability. This also means that professional learning needs to be provided for at least two-years to be effective (Warren and Miller 2013).

Location also plays an integral part to sustainability. It affects teachers in two ways. First, living in remote and very remote locations means that it is difficult and expensive for teachers to leave school and attend professional learning in metropolitan locations. Thus, the professional learning needs to come to them or in locations close to a number of remote schools. Second, resources are difficult to buy in remote and very remote locations. Often teachers need to order resources for their classrooms and have them shipped/flown to their schools. This is timely and expensive. Due to the limited resources in these contexts often the professional learning delivered fails to translate into actions.

A lack of commitment and support for staff from school leadership can also create a barrier for sustainability. In this instance, all staff are not supported to attend professional learning sessions, or continue with professional learning. For example, teacher assistants rarely attend professional learning along side their classroom teachers. In marginalized contexts this is particularly problematic as often the teacher assistants remain at the school once teachers leave or transfer to another school. Thus the knowledge gained from the professional learning is 'lost' from this school, as the personal left at the school have not participated in its creation.

Sustainability has always been at the forefront of RoleM. It was essential to create a sustainable project that would provide students with a stable transition of their mathematics learning through the early years. To prevent such barriers arising RoleM promoted the following conditions: (i) policy changes, that is, embedding the implementation of RoleM into school's policy; (ii) procedural changes, such as ensuring professional learning is ongoing and the RoleM website is utilized, systemizing the organization, storage and access of RoleM materials, providing handovers and PL for new staff, and ensuring that clear links to the Australian mathematics curriculum are maintained; and (iii) support for changes, such as increasing and maintaining school administration support and assisting with appointing a key teacher to support the implementation of RoleM in each school.

An International Perspective of Sustainability

The following presents the New Zealand and Canadian perspectives on sustainability.

New Zealand Perspective by Roberta Hunter and Jodie Hunter[1]
Maintaining the momentum and gaining sustainability is a challenge faced by all mathematics educators and researchers in this space. Given what we have to work with—many of the least experienced teachers and high levels of marginalized students—our challenge begins with making a difference and then it shifts to maintaining the difference. In the *Developing Mathematical Inquiry Communities* (DMIC) project from one year to the next we see staff turn-over of 25 % while at the same time student transience occurs throughout the school year. To address the transient nature of the student population DMIC establishes professional development learning communities across local schools so that as students move from school to school they experience similar mathematics. The year after DMIC project involving professional development is completed schools are helped by establishing a Lesson Study model to cope with teacher attrition. The Lesson Study model allows new teachers to be inducted into the ambitious and culturally responsive pedagogical practices enacted in mathematics classrooms in the schools. The model of 'Teaching as Inquiry' is used while also being embedded within and supported by the core Pāsifika values modeled in the classrooms.

Key implications we have drawn from our New Zealand work which in fact closely match those of RoleM include the need to consider:

- The impact of teacher perceptions and beliefs on the achievement of marginalized learners.
- Professional development needs to focus on growing a combination of content and pedagogical knowledge and this in turn results in enhanced student learning.

[1]This contribution has been written by Associate Professor Roberta Hunter and Dr. Jodie Hunter from Massey University.

- Change is a slow gradual process and teachers need space to do this within their own timeframe although there is a predictable trajectory.
- Sustainability needs to be written into the project and positioned so that the schools can gradually take responsibility for it.

The importance of teachers developing a repertoire of culturally responsive pedagogical actions including considering and working within the core values of the learners they are engaging with cannot be underestimated.

In conclusion, the best outcome for students learning Mathematics at the Margins in New Zealand is the construction of a strong cultural identity and positive mathematical disposition. This is best expressed with quotes to close with from three of our Pāsifika students.

Grace: *When the problems are about us you can see that maths is real and it's useful……not just something random you do at school.*

Sione: *When the maths is about us and our culture it makes me feel normal, and my culture is normal.*

Luana: *Yeah like it is normal to be Samoan or Tongan.*

Canadian Perspective by Jacqueline Ottmann[2]

As the RoleM team addressed the marginality of students, the in-service teacher professional development and resource issues related to mathematics in rural and remote locations, the University of Calgary, Werklund School of Education is striving to meet an educational need for prospective students living in rural and remote Alberta. "Rural Canada is defined as those areas outside urban centres with a population of at least 10,000 (Werklund School of Education 2015). Werklund School of Education offers a unique four-year Bachelor of Education Community-Based program. This blended learning program is specifically designed for prospective students who would prefer to live in their rural and remote locations for the majority of their undergraduate teacher education experience. The students will only be required to take one to two classes on campus each summer. By offering this program, Werklund School of Education is addressing an issue of equity and access to education for people who live in rural and remote communities.

The Community-Based degree program has the following goals:

- Attract students who are interested in pursuing a BEd degree, but are unable to commit to a residency-based program;
- Allow individuals to remain in their local communities for their field placements in which they would serve;
- Help to mitigate the high turnover of teachers in rural and remote areas by hiring those individuals who currently live in those areas.

---

[2]This contribution has been written by Associate Professor Jacqueline Ottmann,University of Calgary Werklund School of Education.

The program has been well received by First Nations communities in Alberta and non-Aboriginal residents in rural and remote areas.

In addition to this, the Werklund School of Education will be committed to supporting students even after they graduate from the Community-Based degree program, and the School is also exploring ways to provide on-going professional learning for in-service teachers who live in rural and remote areas. As in the RoleM study supporting teachers in these unique locations with quality learning, resources and access to 'experts' and mentoring would benefit students' and the whole community learning and overall wellbeing.

Wimmer et al. state,

> "The reality of current demographic patterns in Canadian schools suggests a compelling need in higher education, including teacher education, to become not only better informed about the concerns of Aboriginal peoples but also more responsive to their needs" (2009, p. 821).

By developing teacher preparation programs and professional learning opportunities in and with community consultation, the chances increase for learning to be more meaningful because content and experiences are contextual and culturally relational, flexible and respectful to the 'place and space' where students reside. It is an exciting time in education, a time ripe with possibilities.

Commonalities Between Contexts

When considering the three contexts (Australian, New Zealand, and Canadian), there are a number of commonalities, which appear to contribute to implementing a successful project. These successes also contribute to maintaining the momentum and creating opportunity for sustainability in marginalized contexts.

Across all three contexts:

- the role of community was central to the success of the project. This included consulting with community members to ensure that (a) they supported the project and could see the potential for it in their community, and (b) they provided insight into their own communities so that the project could be adapted for that particular context.
- there was a recognition that teachers are having difficulty in these contexts. This included the understanding that most teachers were beginning teachers with little prior experience working in marginalized contexts.
- both mathematical content knowledge (MCK) and mathematical pedagogical knowledge (PCK) were areas that teachers needed further professional learning. Providing support to build teacher MCK and PCK was essential. Additionally, it was about linking these two areas (MCK and PCK) to culturally appropriate contexts for students.

- blended models of professional learning was offered. Teachers attending professional learning days and then experts working in teachers' classrooms (post-professional learning) to assist them to implement the learnt strategies.
- teacher feedback was central. This provided opportunity to further develop resources and learning experiences so that the mathematics lessons were successful.
- providing appropriate student learning experiences that were culturally tailored for students to encourage engagement in mathematics.
- high expectations and a positive attitude to mathematics was critical for both teachers and students.

Additional Features in the RoleM Project

While there were commonalities across the three contexts, there were additional features in the RoleM project that contributed to sustainable practice and maintaining the momentum.

Role of Leadership in the School

Leaders in the school were central to the sustainability of RoleM. It was the belief that if principals did not 'buy in' to the project, that it would not succeed in the school. The leadership personnel consisted of principals, assistant principals and curriculum leaders. All members were invited to participate in the professional learning days. Leaders who supported their staff, both teachers and Indigenous Teacher Assistants (ITAs), to attend professional learning often had the most success in the project. This success was on three levels. First, sustained engagement and improvement of students' numeracy scores as they transitioned from Foundation to Year 3. Second, changes in the delivery of mathematics with a more consistent approach across the early years. Third, equitable partnerships for teachers and ITAs as they both share the knowledge and learning in the classroom. This leads to a more conducive and engaging environment for student learning by

- embedding the implementation of RoleM into their school's policy and mathematics programs.
- ensuring professional learning is ongoing.
- systemising the organisation, storage and access of mathematical resources and materials.
- providing handovers and PL for new staff.

- supporting teacher change by increasing and maintaining school administration support, appointing a key teacher to support RoleM implementation in the school.
- building relationships with experts in mathematics and using them to support teacher dialogue.

Role of Para-Professionals

As part of ensuring the sustainability of RoleM, it was essential to include the Indigenous Teacher Assistants in the professional learning days. ITAs have a critical role in classroom as they 'walk' and work between the two knowledge systems of Indigenous knowledge and Western knowledge in Australian schools (Fitzgerald 2006). Thus, RoleM funded ITAs to attend professional learning workshops alongside their classroom teacher and leadership staff. It was perceived that this would assist to build collaborative partnerships between teachers, ITAs and the community. Furthermore, by attending the professional learning this would empower ITAs to become more respected and effective in supporting students' learning (Baturo et al. 2007). When teachers and ITAs work in collaboration, Indigenous student learning improves (Malloch 2003). At these sessions the ITAs received numeracy resources that were identical to those given to their teachers. Additionally, due to high levels of teacher turnover in schools, it was essential to include the ITAs in the professional learning (PL). By including the ITAs in PL this would ensure that RoleM could continue once staff had left and new staff had arrived. It also made the learning consistent for students who may have had a change of teacher during the year.

Providing Developed Resources

While the professional learning days focused on improving MCK and PCK, there were also developed resources used on the days and then distributed to teachers. Teachers received a set of RoleM Teachers' Book, which consisted of a book of Learning Experiences and a Blackline Master book (printable/reproducible visual resources). Additionally, hands-on materials were provided to teachers so that once completing the PD they could implement RoleM the next day in their classroom. Underpinning the development of these resources was a recognition that: (1) students learn in a variety of ways; (2) classrooms have students who are at different stages in their learning of mathematics; (3) student engagement is closely associated with student learning; (4) classrooms in disadvantaged contexts are often poorly resourced; (5) teachers often have pre-conceived beliefs that these students are incapable of engaging in the main-stream curriculum; (6) teachers are professionals with an understanding of what works and what does not work in their classrooms. Therefore the materials developed, as part of the RoleM project have 'long shelf life' in classrooms across Australia.

To assist with maintaining the momentum in schools:

- the RoleM materials remain the property of the school and not the classroom teacher therefore ensuring schools always have these materials for incoming new teachers.
- a maintained website has been established with a variety of resources and learning experiences to assist teachers' implementation of the project (www.rolem.com.au). Additionally, this website is maintained so that any curriculum changes can be made to learning experiences and teachers can continue to be supported by RoleM materials.
- a new RoleM test was developed as an online extension of the paper-based tested used in the project. Teachers are able to use these tests to gather diagnostic information about each student and plan for the needs of all of their students (www.exam.rolem.com.au). Additionally, this test can be used to measure a students', or classes', growth across one-three years.

Long-Term Partnerships

Long-term partnerships need to be established with schools and communities in order for a project to become sustainable. Too often teachers experience a once off professional learning day, which is not implemented into practice. By establishing a long-term partnership with the school, teachers have more opportunity to learn and implement the MCK and PCK they are being exposed through the professional learning. Additionally, it ensures success for long-term goals. It is difficult to implement change in a school, across a number of year levels, without on-going support and partnerships from experts. Experts establishing a presence in the school and community demonstrates they are committed to the staff and students. Additionally, this provides an opportunity for the learning experiences to be adapted and refined for the needs of each class.

Actions that Make a Difference

In summary to the book, the following section will present actions that make a difference according to professional learning, teaching mathematics, and a whole-school approach to implementing change.

Professional Learning

Actions that make a difference to professional learning incorporate:

- Professional learning that focuses on learning experiences, developmental sequences, and catering for diversity in mathematics.
- Professional learning days that afford teachers the chance to trial ideas and reflect with experts in the field of mathematics.
- Provision of research-based materials for classroom trials.
- Site visits by experts in mathematics teaching and learning incorporating modelling in teachers' classrooms.
- Ongoing sequences of professional learning that are responsive to teachers' needs.

At least two years of professional learning are required for substantive teacher change to occur (Warren and Miller 2013). Professional learning days are not enough. In class modelling by experts in the field is a critical component to supporting this change.

Actions that make a difference to teaching mathematics incorporate:

- representing mathematical concepts in multiple ways.
- using students' engagement and assessment to inform teaching.
- implementing tailored learning experiences that cater for the participation of *all* learners including those experiencing difficulties and those who need extension.
- building on students' strengths and their cultural backgrounds.
- encouraging students to orally communicate their mathematical understanding.
- recognising that *all* students are capable of learning mathematics.

Finally, Indigenous Teacher Assistants are often key members within their community and provide the liaison between school and community. Thus, the strengthening of partnerships between teachers and ITAs not only assists the sustainability of initiatives but also helps to bridge the connection between school and community.

Conclusion

This book provides a positive story for students (and their teachers) living in marginalized contexts. It demonstrated that through ongoing commitment to teaching and learning, students are capable of achieving at aged appropriate levels in mathematics. It is our belief that strong foundations in mathematics in the early years lead to stronger outcomes in the later years of schooling.

The focus of this book was to share the findings from a four-year longitudinal study *Representations Oral Language and Engagement in Mathematics* (*RoleM*) that was situated in the most marginalized schools in Queensland, Australia. The overarching aim of the RoleM project was to improve marginalized students' numeracy. This was achieved by:

- delivering quality teacher professional learning with regard to planning and implement effective mathematical strategies in conjunction with examining the assessment of students' growth in mathematics.
- facilitating collaborative partnerships between teachers, Indigenous teacher assistants and the community.
- empowering ITAs to become more respected and effective in supporting students' learning.
- creating and implementing culturally appropriate learning activities and resources.
- ensuring that the improved student learning of mathematics was sustainable.
- engaging leadership staff in the project as to ensure sustainability.

Sustainability is an important dimension in making a difference for students in marginalized contexts. By making a project sustainable, it ensures something has been left behind once the researchers have gone.

In this book we have shared the journey of the teachers and students as they transition to successful learners of mathematics. We strongly believe that at the essence of programs such as RoleM the students are central. It is about making a difference to students and their learning that may later impact on their life choices. Teachers and teacher assistants support students in this journey as they achieve their true potential. All teachers and students in this project were successful either through improved positive dispositions to mathematics, or increase content or pedagogical knowledge of mathematics. While this story is the end of the first chapter for RoleM, there are more questions and journeys to explore as this research continues into new areas and levels of schooling.

References

Anderson, S. E., & Stiegelbauer, S. (1994). Institutionalization and renewal in a restructured secondary school. *School Organisation, 14*(3), 279–293.

Baturo, A., Cooper, T., & Doyle, K. (2007) Authority and esteem effects of enhancing remote Indigenous teacher-assistants' mathematics-education knowledge and skills. In J. Woo, H. Lew, K. Park, & D. Seo (Eds.) *Proceedings 31st Annual Conference of the International Group for the Psychology of Mathematics Education* (pp. 57–64). Seoul, Korea: MERGA.

Earl, L., Watson, N., Levin, B., Leithwood, K., Fullan, M., Torrance, N., et al. (2003). *Watching and learning 3: Final report of the external evaluation of England's National Literacy and Numeracy Strategies*. London, UK: DFES.

Fink, D. (2000). *Good school/real school: The life cycle of an innovative school*. New York: Teachers' College Press.

Fitzgerald, T. (2006). Walking between two worlds: Indigenous women and educational leadership. *Educational Management Administration and Leadership, 34*(2), 201–213.

Hargreaves, A., & Fink, D. (2003). The seven principles of sustainable leadership. *Educational Leadership, 61*(7), 8–13.

Knight, N. (2005). The contested notion of sustainability: Possibility or pipe dream for numeracy reforms in New Zealand. In P. Clarkson, A. Downton, D. Gronn, M. Horne, A. McDonough, R. Pierce, et al. (Eds), *Building connections; Theory, research and practice: Proceedings of the 28th annual conference of the Mathematics Education Research Group of Australasia* (MERGA)(pp. 467–474). Auckland: MERGA.

Malloch, A. (2003). Teacher and tutor complementation during an in-class tutorial program for Indigenous students in a primary school. In S. McGinty (Ed.) Sharing success an Indigenous perspective. *Papers from the second National Australian Indigenous education Conference.* (pp. 181–204). Townsville, Queensland

Millett, A., Brown, M., & Askew, M. (2004). Drawing conclusions. In A. Millett, M. Brown, & M. Askew (Eds.), *Primary mathematics and the developing professional* (pp. 245–256). Dordrecht, The Netherlands: Kluwer.

Timperley, H. (2008). *Teacher professional learning and development.* Retrieved from http://www.orientation94.org/uploaded/MakalatPdf/Manchurat/EdPractices_18.pdf

Timperley, H., Wilson, A., Barrar, H., & Fung, I. (2007). *Teacher professional learning and development: Best evidence synthesis iteration [BES].* Wellington, NZ: Ministry of Education.

Warren, E., & Miller, J. (2013). Enriching the professional learning of early years teachers in disadvantaged contexts: The impact of quality resources and quality professional learning. *Australian Journal of Teacher Education, 38*(7), 91–111.

Warren, E., & Quine, J. (2013). Enhancing teacher professional development for early years mathematics teachers working in disadvantaged contexts. In L. English & J. Mulligan (Eds.), *Reconceptualizing early mathematics learning: Advanced series in mathematics education* (pp. 283–308). Berlin, Heidelberg: Springer.

Werklund School of Education. (2015). *4 year community-based Bed pathway*. Calgary: University of Calgary. Retrieved from http://werklund.ucalgary.ca/upe/prospective-students/bed-program/4-year-bachelor-education/4-year-communitybased-bed-pathway

Wimmer, R., Legare, L., Arcand, Y., & Cottrell, M. (2009). Experiences of beginning Aboriginal teachers in band-controlled schools. *Canadian Journal of Education*, *32*(4), 817–849. Retrieved from http://files.eric.ed.gov/fulltext/EJ883523.pdf